FEYNMAN WAS GAZING AT A RAINBOW...

as if he had never seen one before. Or maybe as if it might be his last.

I approached him cautiously and joined him in staring at the rainbow. It wasn't something I normally did—in those days.

"Do you know who first explained the true origin of the rainbow?" I asked.

"It was Descartes," he said. After a moment he looked me in the eye. "And what do you think was the salient feature of the rainbow that inspired Descartes's mathematical analysis?" he asked.

"I give up. What would you say inspired his theory?"

"I would say his inspiration was that he thought rainbows were beautiful..."

—from FEYNMAN'S RAINBOW

ACCLAIM FOR *FEYNMAN'S RAINBOW*

"An exhilarating book...one that reflects the radiance of its subject and so warms even as it instructs."

—David Berlinski, author of *A Tour of the Calculus*

"An inspirational rite-of-passage book that provides a peek into the world of academic scientists as they struggle to understand nature and life."

—*Tampa Tribune*

"An inspiring and very readable portrait of a free-spirited genius."

—*Kirkus Reviews*

"FEYNMAN'S RAINBOW is not a book that provides answers, and yet it is a comforting book. How many people do we know for whom physics conjures painful high-school memories? Have they got an hour to spare? For them, this book would be a surprisingly rewarding gift."

—*Toronto Globe and Mail*

more...

"Like their celebrated quarks, the lives of scientists are strongly confined and shaped by the interplay of 'truth,' 'beauty,' and 'strangeness.' FEYNMAN'S RAINBOW offers a rare glimpse into this fascinating world. I enjoyed every page of it."

—Fritjof Capra, author of *The Tao of Physics*

"Thanks to Mlodinow and his intimate, engagingly written book, readers will appreciate Feynman for the great man and scientist he was."

—*Jerusalem Post*

"More than a memoir of friendship, this book offers a unique glimpse into Feynman's final years."

—*Seed*

"An entertaining look at...creativity, science, and life."

—*Body & Soul*

"Effortlessly charming...FEYNMAN'S RAINBOW moves gracefully and with a tender mixture of respect and affection for a man who shone with a child's sense of wonder, and who was one of the twentieth century's great scientific minds."

—Wigglefish.com

"Offers priceless insights into the culture of science—the cranky, eccentric, very human people who do it, the quixotic way that it is done, and the extraordinary passion that is required to do it well....As an added bonus, it is written so well that you can take it to the beach."

—*Research News and Opportunities in Science and Theology*

Also by Leonard Mlodinow

Euclid's Window: The Story of Geometry from Parallel Lines to Hyperspace

Feynman's Rainbow

A
Search for
Beauty
in Physics
and in Life

Leonard Mlodinow

Warner Books

New York Boston

Warner Books

Time Warner Book Group
1271 Avenue of the Americas, New York, NY 10020
Visit our Web site at www.twbookmark.com

Printed in the United States of America
Originally published in hardcover by Warner Books
First Trade Printing: May 2004
10 9 8 7 6 5 4

The Library of Congress has cataloged the hardcover edition as follows:
Mlodinow, Leonard.
Feynman's rainbow : a search for beauty in physics and in life / Leonard Mlodinow.
p. cm.
ISBN: 0-446-53045-X
1. Mlodinow, Leonard, 1954- 2. Feynman, Richard Phillips. 3. Physicists—United States—Biography. I. Title.
QC16.M635 A3 2003
530'.092—dc21
[B] 2002031137

ISBN: 0-446-69251-4 (pbk.)

Book design by Giorgetta Bell McRee
Cover design by Jackie Merri Meyer
Cover illustration by R.O. Blechman

To Donna Scott

Acknowledgements

I am grateful to Jamie Raab at Warner Books for seeing the promise of this book, and to Les Pockell and Colin Fox, my editors at Warner, for their invaluable support and insightful suggestions, not to mention all their hard work; to Susan Ginsburg for her guidance, encouragement, friendship, and—most of all—her faith in me; to Michelle Feynman, Eric Wilson, Mark Hillery, Matt Costello, Erhard Seiler, Fred Rose, Annie Leuenberger and Stephen Morrow for their input, support and friendship; to Donna Scott, for her love and friendship; and to the Five Spot bar in Brooklyn, where I was always kindly tolerated as I lingered over a few beers, pondering the meaning of physics and life.

So spoke an honest man; the outstanding intuitionist of our age and a prime example of what may lie in store for anyone who dares to follow the beat of a different drum.

—Nobel Laureate Julian Schwinger,
in his obituary of Feynman in
Physics Today,
February 1989

PREFACE

Fewer than eight hundred Americans earn a Ph.D. in physics each year. Worldwide, the number is probably in the thousands. And yet from this small pool comes the discovery and innovation that shapes the way we live and think. From X-rays, lasers, radio waves, transistors, atomic energy—and atomic weapons—to our view of space and time, and the nature of the universe, all this has arisen from this dedicated pool of individuals. To be a physicist is to have an enormous potential to change the world. It is also to share a proud history and tradition.

To a physicist, the most important years are those of graduate school and immediately after. It is the time you find yourself and build your career. This book is about my time just after graduation in 1981, when I was on the faculty of the California Institute of Technology, one of the world's top research facilities.

My experience there was not the usual one. I arrived at Caltech feeling lost and intimidated. I was uncertain of my abilities and unusually unfocused in the vision I had for my future. I was also unusually lucky to have landed an office just down the hall from one of the greatest physicists of the century—Richard Feynman. It was Feynman who, while on the 1986 space shuttle commission, made worldwide headlines demonstrating the solution to the riddle of the failed O-ring by dunking it in ice water and pounding it on the table to show it had become brittle. That was vintage Feynman: a triumph of common sense over computer models, of insight over equations. A year earlier Feynman's irresistible memoir, *"Surely You're Joking, Mr. Feynman!"* had exploded on the bestseller lists. In the popular psyche, Feynman has become, since his death in 1988, the Einstein of modern times. In 1981, Feynman was largely unknown outside the physics world, though within it he had been a legend for decades.

I had been given my fellowship because my Ph.D. thesis, which was on quantum theory in infinite dimensions, had caught the attention of some notable physicists. Did I really fit in here, with two Nobel Prize winners down the hall, and the best students in the country all around me? Week after week I came to my office and pondered the great open problems of physics. No ideas came to me. I was certain that my earlier work had been a fluke and that I would never again discover anything worthwhile. I suddenly understood why Caltech had one of the highest suicide rates of any college in the country.

One day I got the courage to knock on the door to Feynman's office and, to my surprise, found that I was welcome. He had just undergone his second surgery for the cancer that would eventually kill him. Over the next two years we spoke many times, and I had the opportunity to ask him questions, such as: How do I know if I have what it takes? How does a scientist think? What is the nature of creativity? From this famous scientist near the end of his days, I found the answers I sought about the nature of science and the scientist. But more than that, I discovered a new approach to life.

This book tells the story of my first year on the Caltech faculty, beginning in the winter of 1981. In that sense it is the narrative of a young physicist trying to find his place in the world, and of the famous, old, and dying physicist whose wisdom helped him. But it is also the story of Richard Feynman's last years, his rivalry with fellow Nobel laureate Murray Gell-Mann, and the beginnings of string theory, today the leading theory on the frontiers of physics and cosmology.

This book tells a story, but it is not a novel. I took notes on and recorded many of my conversations with Feynman because I was awestruck. The passages in italics are based on these notes and the transcripts of some of these discussions. Everything I describe in this book happened to me. But I have combined and altered events, and, other than the historical figures and those whose specific work I quote—Feynman, Murray Gell-Mann, Helen Tuck, John Schwarz, Mark Hillery, and Nick Papanicolaou—I have altered names and personalities in order to best portray my experience.

I am grateful to Caltech for being such a lively and

exciting place to do research, and for, so long ago, having the confidence that they had in me; and I am especially grateful to the late Richard Feynman, for his many lessons on life.

Feynman's Rainbow

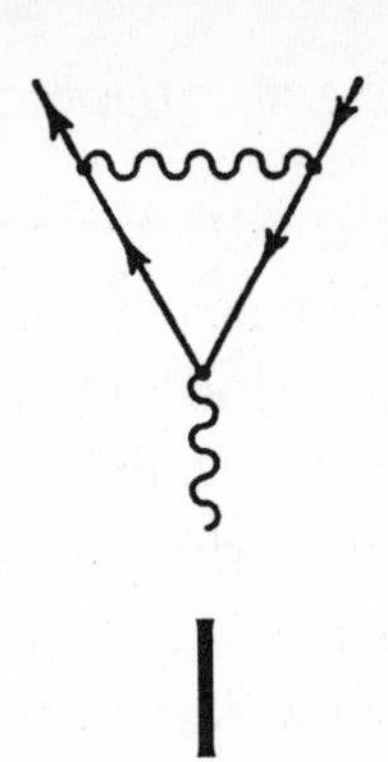

I

IN A GRAY CEMENT building on the olive tree–lined Caltech campus on California Boulevard in Pasadena, a thin man with longish hair steps into his modest office. Some students, on this planet less than one-third as long as the professor has been, stop in the hallway and stare. No one would say a word if he didn't come to the office this day, but nothing could keep him away, especially not the surgery, the effects of which he would no longer allow to ruin his routine.

Outside, bright sun bathes the palm trees, but it is no longer the withering sun of the summer. The hills rise, brown now giving way to green, their vegetation reborn with the coming of the more hospitable winter season. The professor might have wondered how many more cycles of green and brown he would live to witness; he knew he had a disease that would kill him. He loved life, but he believed in natural law, and not in miracles. When his rare form of cancer was first discovered in the summer of 1978, he had searched the literature. Five-year survival rates were generally reported to be less than 10 percent. Virtually no one survived ten years. He was into his fourth.

Some forty years earlier, when he was almost as young as the students currently around him, he had sent a series of papers to the prestigious journal *Physical Review.* The papers contained odd little diagrams, which constituted a new way of thinking about quantum mechanics, less formal than the standard mathematical language of physics. Though few seemed convinced of his new approach, he thought how amusing it would be if some day that journal would be full of his diagrams. As it turned out, the method they reflected proved to be not only correct and useful, but revolutionary, and on that day late in 1981, in the *Physical Review*, his diagrams were ubiquitous. They were about as famous as diagrams get. And he was about as famous, at least in the world of science, as scientists get.

The professor has been working on a new problem the past couple of years. The method he worked out in his student days had been wildly successful when applied to a theory called quantum electrodynamics. That is the theory of the electromagnetic force that governs, among other things, the behavior of the electrons that orbit the nucleus of the atom. These electrons impart to atoms their chemical properties and their spectral properties (the colors of light they emit and absorb). Hence the study of these particular electrons and their behavior is called atomic physics. But since the professor's student days physicists had made great progress in a new field called nuclear physics. Nuclear physics looks beyond the electronic structure of atoms to the potentially much more violent interactions of the protons and neutrons within the nucleus. Though protons are subject to the same electromagnetic force that governs the behavior of the atomic electrons, these interactions are dominated by a new force, a force that is far stronger than the electromagnetic force. It is called, fittingly, the "strong force."

To describe the strong force a grand new theory had been invented. The new theory had some mathematical similarities to quantum electrodynamics, and it was given a name that reflected these similarities—quantum chromodynamics (despite the root, *chromo*, it has nothing to do with color as we know it). In principle quantum chromodynamics provided a precise quantitative description of protons, neutrons, and related particles and how they interact—how they might bind to each other, or behave in collisions. But how do we extract descriptions of these processes from the theory? The professor's approach applied in principle to this new theory but practical complications arose. Though quantum chromodynamics had had certain triumphs, for many situations neither the professor nor anyone else knew how to use his diagrams—or any other method—to extract accurate numerical predictions from the theory. Theorists couldn't even calculate the mass of the proton—a very basic quantity that had long ago been accurately measured by the experimentalists.

The professor thinks, perhaps, that with the months or years he has left on earth he'll play around with the problem of quantum chromodynamics, considered one of the most important of its day. To create the energy and will he needs for his effort, he tells himself that everyone else who had for so many years unsuccessfully attacked this problem lacked certain qualities that he possesses. What they are he, Richard Feynman, isn't sure: an oddball approach, perhaps. Whatever those qualities are, they had served him well—he had one Nobel Prize, but might arguably have deserved two or three when you considered all the wide-ranging and important breakthroughs he had made in his career.

Meanwhile, in 1980, several hundred miles north in Berkeley, a much younger man had sent off a couple papers with his own new approach to solving some of the old mysteries of atomic physics. His method offered answers to some difficult problems, but there was a catch. The world he explored in his imagination was one in which space has an infinite number of dimensions. It is a world with not just up/down, right/left, and forward/backward, but also a countless array of other directions. Could you really say anything useful about our three-dimensional existence by studying a universe like that? And could the method be extended to other areas of study, such as the more modern field of nuclear physics? It would turn out that it is promising enough that this student received a beginning faculty appointment at Caltech, and an office down the hall from Feynman.

The night after receiving that offer of employment, I remembered lying in my bed half my life earlier, wondering what it would be like the next day, my first day in junior high. More than anything else, as I recall, I was worried about gym and showering in front of all those other boys. What I was really worried about was ridicule. I would be exposed, too, at Caltech. In Pasadena there would be no faculty advisor, no mentor, just my own answers to the hardest problems the best physicists could think of. To me, a physicist who didn't produce brilliant ideas was one of the living dead. At a place like Caltech, he would also be shunned, and soon unemployed.

Did I have it or didn't I? Or was I asking the wrong question? I started talking to the thin, dying professor with long hair in an office down the hall. What the old man told me is the subject of this book.

II

THE STORY REALLY BEGINS in the winter of 1973. I lived on a kibbutz, a communal farm, in Israel, in the foothills near Jerusalem. My hair was shoulder-length, and my politics pacifist, but I was there because of a war, the Yom Kippur War, named after the day on which it had started. Though it was mostly over by the time I arrived, its vestige was dragging on. The troops were still mobilized. This led to a serious labor shortage. I took leave from college in the midst of my second year to go help out.

I was twenty and felt grown-up. But I was still a child—guided, taken care of, and protected. The kibbutz experience was my first experience in many areas of life—my first time in a foreign country, my first time working with farm animals, my first time taking refuge in a bomb shelter while shells exploded outside. And it was the first time I ever lived without certain amenities we take for granted—stereos, televisions, telephones . . . indoor bathrooms.

At night there was little to do except chat with the

other volunteers, look at the stars, or visit the small "library" on the kibbutz, which had a few dozen books in English. A number of the books in the library were physics books, apparently donations by a kibbutznik who had attended college in the States. I had a double major at the time—in chemistry and mathematics—and everyone who knew me assumed I'd someday be a chemistry professor at a major university. I'd always been an academic kid, and as long as anyone could remember, my two subjects were chemistry and math. The "advanced" physics course I had in high school had been dry and boring. I didn't get the big fuss everyone made over Isaac Newton—who could get excited about the speed of a ball rolling down an inclined plane, or the force of a weight you dropped from the second floor? It was no competition to the fireworks and rockets I could throw together in a chemistry lab or the curved space I could dream about in math courses. Still, given the thin set of choices, I eventually looked over the physics books.

One of them was a paperback called *The Character of Physical Law* by a fellow I had vaguely heard of—Richard Feynman. The book was the transcript of some lectures he had given in the sixties. I picked it up. It explained, without employing mathematics, the principles of modern physics, especially quantum theory.

"Quantum theory" is not really a particular theory, but rather a type of theory. A quantum theory is any theory based on the "quantum hypothesis," revealed to the world by Max Planck in the year 1900, which states that certain quantities such as your energy can take on only certain discrete values. For instance, at any given height above the surface of the earth, you possess something called gravitational potential energy. This is the energy with which you'd hit the ground if you fell from that height (in the absence of air resistance). In a quantum theory of gravity, your gravitational potential energy could not have just any value—there would be only a discrete set of energies you could possess. There is even a minimum possible energy that corresponds to being a little above the earth's surface. This has recently been measured in an experiment on neutrons, for which the minimum energy corresponds to a height of roughly five ten-thousandths of an inch. If your ruler has the accustomed precision, it is a restriction you'd hardly be able to detect. Quantum effects are important, however, when you study objects like neutrons, nuclei, or atoms.

Theories that do not incorporate Planck's quantum hypothesis are called classical theories. Obviously, before 1900, all theories in physics were classical theories. For the most part classical theories work just fine unless you are concerned with the nuances of behavior on the atomic scale, or smaller. This proved to be most physicists' focus for most of the next hundred years.

Physicists spent the first few decades of the twentieth century working out the consequences of Planck's quantum hypothesis. One of them is the famous uncertainty principle, which states that in nature there are certain pairs of quantities whose values cannot be simultaneously pinpointed. For instance, if you determine the position of an object with great precision, then you cannot know its velocity very precisely. Again, for the large objects we encounter in everyday life, these limitations are not noticeable, but for the constituents of atoms, they make an enormous difference.

Another consequence of quantum theory is what physicists call "wave-particle duality," which means that, under certain circumstances, particles such as electrons exhibit the behavior of waves, and vice versa. For example, if you shoot a series of electrons at a tiny slit in a wall, as they pass through they will spread in a circular pattern like a water wave that passes through a small opening. And if you put two tiny slits in the wall, you will see interference ripples similar to those you see when two water waves collide. An electron as a wave is an electron spread out in space, an electron that acts as if it were an excitation of some pervasive medium rather than a discrete object in itself. On the other hand, wave-particle duality also tells us that there are circumstances in which waves of energy exhibit particlelike behavior. An example of this is light. We have known light through the ages mostly as a wavelike phenomenon. For instance, think of the way it bends as it passes through a lens, or the way it spreads out in a prism. But it can also behave as a particle, a discrete localized object, which we call a photon. This concept of light proved to be the key to understanding the photoelectric effect, in which certain metals eject an electron after being impinged upon by a photon. Einstein, the first to accept the quantum hypothesis as a fundamental physical law, explained certain mysterious properties of the photoelectric effect in these terms in one of his famous papers of 1905. (It was for this work, not his controversial relativity theories, that he received the 1921 Nobel Prize.)

Today, we have quantum versions of the old classical theories, such as quantum electrodynamics, and we also have new quantum theories describing forces not even known in Planck's day, such as quantum chromodynamics. But there is one exception to this trend of quantum-ization: the theory of gravity. No one has ever figured out how to incorporate the quantum hypothesis into Einstein's theory of gravity, called general relativity.

Quantum mechanics makes for a fascinating world. I was naturally curious about it, but I had always found textbook descriptions dry and technical. Feynman made it wondrous and magical. I was riveted. I wanted to read more.

There were three other books by Feynman in the collection—his three-volume *The Feynman Lectures on Physics,* from a college survey course he gave at Caltech. They included a picture of the author—an action shot of a happy fellow playing bongo drums. Those books were unlike any textbooks I'd ever seen. They were chatty; they were amusing; it felt as if Feynman was in the room speaking to you. The discussion of mechanics spoke of Newton, but also Dennis the Menace. The section on the kinetic theory of gases included questions like, "Why do we deal with this subject now at all?" The chapters on light included a digression on "some very interesting things (which) have been discovered about the vision of the bee." But Feynman didn't just make physics sound fascinating. Without ever saying so, he also made it sound important. As if a physicist, with an idea, could singlehandedly change the world, and the way people view it. I found myself thinking over problems and issues from Feynman's books as I drove the tractor hauling chicken eggs, herded cattle, or peeled potatoes in the communal kitchen.

By the time I landed back home in Chicago that summer, I had decided I wanted to study physics.

In view of its great impact on me, the kibbutz had allowed me to keep *The Character of Physical Law,* in exchange for a pair of old blue jeans. Toward the end of Feynman's book, I underlined a passage: "We are very lucky to live in an age in which we are still making discoveries. It is like the discovery of America—you only discover it once. The age in which we live is the age in which we are discovering the fundamental laws of nature, and that day will never come again." I promised myself that I would someday make a discovery. And that I would someday meet this Professor Feynman.

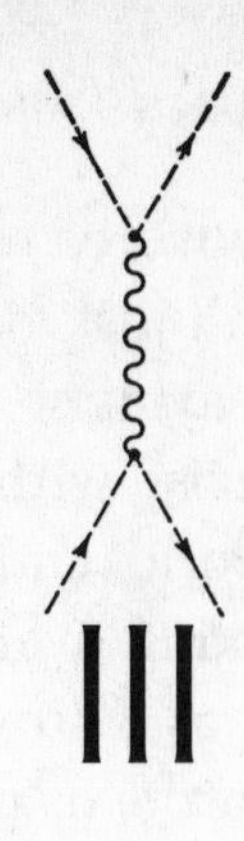

III

FALL 1981. A lot had happened since my days in Israel. I had added a major in physics, graduated, gone to graduate school at Berkeley, and obtained my Ph.D. My parents came to the graduation. It was the last major event of my life in which we would be together as a family. It was the end of my childhood.

Due to some formalities regarding my dissertation—namely, writing it—I arrived at Caltech well after the start of the term. As a private college Caltech had escaped the budget cuts Ronald Reagan had imposed on state schools, especially Berkeley, before he himself graduated from governor to president. Caltech enjoyed one of the highest per capita endowments of any college in the nation. It showed. The campus was beautiful, and serene. And it was large, considering that Caltech undergraduates numbered only in the hundreds. Most of it lay on a site several blocks on each side that was not intruded upon by city streets. Instead, broad sidewalks punctuated by well-kept lawns, shrubbery, and craggy gray olive trees wound their way amongst the low buildings, many of Mediterranean-style architecture. It was a place to feel peaceful and protected, free to forget the outside world and focus on pursuing your ideas.

I felt that to have a job—any job—in academic physics was a privilege. People sometimes scoffed at academia because of the relatively low pay. But I had seen too many "adults" work too many hours at jobs they did not like in order to amass things they only thought they needed, and then, decades later, regret their "wasted" years. And I had seen my father work long, arduous hours just to make ends meet. I had vowed to have a better life than that. The most valuable asset I figured I could earn was the ability to spend my time doing something I liked.

At first I was ecstatic that not only did I have an academic job, but it was at an elite university—the home of my hero Feynman. And it was a dream job, a particularly prestigious multiyear fellowship with complete academic freedom. But as my start date came nearer, the ecstasy dissolved and a strange thought began to crystallize: These people at Caltech might actually expect something of me. Before my dissertation had been officially accepted, I was just a promising student. My assignment was to ask questions, learn, and make the naïve mistakes that cause professors to smile and remember their own carefree days of youth. Now suddenly I was on the faculty. Students would be coming to me for wisdom. Famous professors would mutter something at the water cooler and expect an intelligent reply. Editors at prestigious physics journals would be holding spots open for articles describing my latest momentous discovery.

I formulated a strategy to keep the pressure off: Keep expectations low, stay unnoticed, and, I hoped, discover that except for a couple of Feynman types, everybody at Caltech was as ordinary as I was.

On my first day, I was called into the department chairman's office. At Caltech they grouped the departments of physics, mathematics, and astronomy into one division, so this guy was really the head of all three departments. I didn't see why a person so high up would need to see a guy like me. All I could think of was that I was being called in because they realized they had given me the fellowship by mistake. *I'm sorry,* I pictured him saying, *my secretary sent the offer letter to the wrong guy. We meant to hire a fellow named Leonard M. Lodinow, not Leonard Mlodinow. You must know of him, Dr. Lodinow from Harvard? Anyway, it was an easy mistake, you must admit.* In my daydream, I admitted it, and started looking for another job.

When I got to the chairman's office I found a middle-aged man, balding, holding a cigarette in his fingers. I later heard he had ulcers. He smiled, stood, and waved me in. The smoke from his cigarette left a wispy trail in the air. He spoke with an authoritative voice and a German accent.

"Dr. Mlodinow, welcome. Things all finished up in Berkeley? We've been looking forward to your arrival." We shook hands and took seats.

I knew his comment was meant to be encouraging, but having the head of physics, mathematics, and astronomy personally looking forward to my arrival didn't quite fit my strategy of lying low. On the other hand, at least he wasn't telling me the fellowship was a mistake. I tried to act pleasant as my stomach tightened even further.

"How do you like southern California so far?" He leaned back in his chair.

"I haven't seen much of it yet," I said.

"Of course not. You just got here. How about the campus? Been to the Athenaeum yet?"

"I had lunch there today." Actually for me it had been breakfast. I worked late and slept late in those days.

The Athenaeum was the faculty club, a fifty-year-old building done in what I was told was the campus's "Spanish Renaissance style." Inside, there was a lot of fine wood, velvet curtains, and elaborately painted ceilings. I heard that upstairs there were a few guestrooms. I thought it felt like a fine resort, but I wasn't sure, never having been to a fine resort.

"Did you know Einstein stayed there for two winters before he settled in Princeton?"

I shook my head.

"Some say he only settled in Princeton because we refused to give a staff position to his assistant. If I had been around we wouldn't have made that mistake." He chuckled.

We made a little small talk. His secretary came in with a phone message, and he told her no messages until we were finished. He studied me for a moment.

"Let me guess. You're wondering what you're doing here?"

Did he see right through me?

"I guess because people liked my graduate work?"

"No, not here at Caltech. Here in my office."

"Oh . . . actually, yeah, I was wondering . . ."

"I'll tell you why. I asked you here because you have a special position at Caltech, and because Caltech is a special place. That means that you deserve a special welcome, a personal welcome, from me."

To someone else his welcome might have sounded like a friendly gesture. But I couldn't help thinking there was an implied *and remember, just in case we're wrong, I'll be watching* at the end of his sentence.

"Oh . . ." I muttered. "Thanks."

He took a hit from the cigarette and leaned back in his chair.

"How much do you know about Caltech?" he said.

I shrugged. "I know the physics department."

"Of course, and just down the hall from your office, as I'm sure you've noticed, are the twin titans of physics, Dick Feynman and Murray Gell-Mann."

Actually this was news to me. I hadn't yet been shown my office.

"But you'll find as you explore our campus further that Caltech has a rich history you might not be fully aware of. Oh, you probably know that it was here that Linus Pauling discovered the nature of the chemical bond. But did you know that it was at Caltech that Charles Richter and Beno Gutenberg invented the Richter scale? Or where computer pioneer Gordon Moore received his Ph.D.?"

"No, I didn't."

"It was. And I'm sure that, as a physicist, you know that it was here that antimatter was discovered. But you might not know it was at Caltech where the principles of modern aviation were conceived, and where the age of the earth was first accurately determined. Or that it was here where Roger Sperry figured out that the right and left hemispheres of the brain have different functions—left for language, right for visual and spatial functions. It was also at Caltech that molecular biology was practically invented. One of the key people in that was Max Delbrück, a physicist like yourself. For this, in 1969, he got the Nobel Prize."

He chuckled again. I didn't see any humor in the conversation, but I tried to chuckle back.

"Do you know how many Nobel Prizes have been received by members of the Caltech community?"

I shook my head. I'd never thought about it.

"Nineteen. By comparison, MIT, which is roughly five times our size, boasts only twenty."

I wondered if they also kept track of how many members of the Caltech community were dismal failures.

"Why am I telling you this? Because even as we speak, the great triumphs of the future are going on today. Explore. Learn what people are doing. You'll be surprised—and, I hope, stimulated. Beginning today, you, too, are a part of our great intellectual tradition."

If I had felt the least bit comfortable before, this trip down the memory lane of genius had definitely made me carsick. I wanted to tell him it sounded as if I had six months to prove myself, and then it would be all over. But I didn't think this was quite the right time and place to open up. So what I said was: "I'll try to live up to it."

He accepted my hopeless wish with great enthusiasm. "Oh, we think you will! That's why we offered you the fellowship we did. Most postdoctoral fellows come here to work under the supervision of a specific professor. Not you. You, Dr. Mlodinow, are a free agent. You are accountable to no one but yourself. You may choose to teach if you wish, which most postdocs cannot, or you may choose not to teach. You may conduct research in physics, or, like Max Delbrück, in biology, or in any other field you wish. If you want, you can use your time to design sailboats! It is all up to you! We give you this freedom because we have judged you to be the best of the best, and we have confidence that, given the freedom, you will do great things."

His pep talk was heartfelt, and he was good at it. But I was the wrong subject. I left his office feeling as I did in a dream I once had. I was in an elevator going up, on my way to my office at Berkeley, when I suddenly realized that I was naked—I had forgotten to put my clothes on that morning. So there I was with a choice: push the stop button, which would delay my having to get out, but set off the alarm and call attention to myself. Or wait for the door to open and try to get to my desk without anyone noticing. In life, as in my dream, I chose the latter.

Some days later I was in my office pondering my plight when I was suddenly offered the opportunity to deaden my nerves with champagne. The whole campus was celebrating, as it had been announced that for his split-brain research Roger Sperry had just won the 1981 Nobel Prize in Physiology or Medicine. In Nobel laureates, Caltech and MIT were now tied. One hemisphere of my brain was proud and excited to be in the midst of this, the other disquieted, as if the pressure had just been turned up a notch.

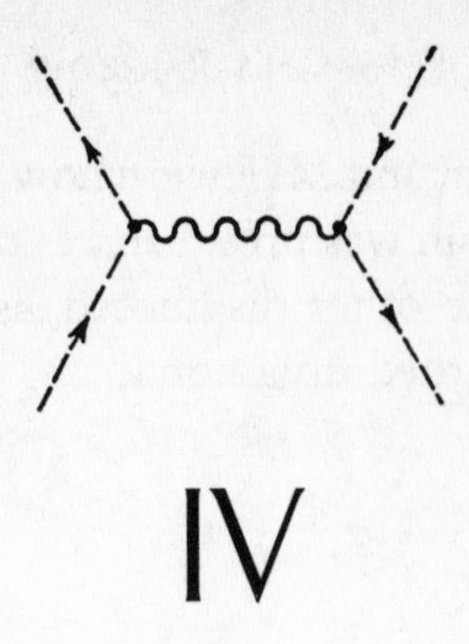

IV

WHEN I WAS FINALLY shown my office, it turned out to be right next door to that of Murray Gell-Mann—one of the twin titans the department chair had mentioned. Some days later I introduced myself and we talked a little over the tea and cookies table that was where people went after a seminar. Murray looked exactly as I had expected from the pictures I had seen—right down to his trademark bolo tie. I told him my name. He didn't tell me his—when you are that famous why bother—but he did repeat mine. It was unrecognizable to me, but, he told me, it was the "correct" (Russian) pronunciation. He also gave the etymology. I didn't ask him the origin of his own unusual name; it turned out the hyphen was his father's invention. Almost everyone called him by his first name, anyway. Feynman was "Dick" to a much smaller group.

Murray's ideas had dominated physics for over twenty years, but his most famous achievement was to invent, in the 1960s, an elegant mathematical system for classifying and explaining the properties of the dozens of known subnuclear particles. Except for the more traditional nuclear components—protons and neutrons—these subnuclear particles decayed within a small fraction of a second, and had been discovered in the last few decades. They came into existence only when, for instance, protons were smashed together. To account for the mathematical order he found amongst this zoo of subnuclear particles, Murray later proposed that the proton, neutron, and the other particles had an internal structure, being made of different combinations of just a few more basic building blocks. These were "sub-subnuclear particles"—that is, particles within the particles that make up the nucleus. He dubbed them quarks. Individual quarks have never been seen, but eventually physicists grew to accept Murray's theory. This has earned him comparisons with Dmitri Mendeleev, who invented the Periodic Table of Elements. Like Murray's system, the Periodic Table organizes the chemical elements into groups based on common properties. And like Murray's system, this order amongst the elements was eventually explained in terms of an internal structure—in this case, the atom's internal structure of particles later named electrons.

Murray's work won him a Nobel Prize and helped him to become one of the most influential scientists of the postwar era. Yet he appeared to have an inferiority complex and seemed anxious to show off how brilliant he was. It didn't matter if you talked about particle accelerators or septic tanks, he could and would tell you how they work, the crucial specs, and what to look for in the latest model. His conspicuously "correct" pronunciation of my name was no aberration; he seemed to seek opportunities to say foreign words, such as the names of cities, so that he could show off his ability to pronounce them like a native. One moment you'd be listening to a seemingly normal native New Yorker, then suddenly his face would contort and for the next few words he would be Quebecois, or Russian, or Chinese. Once, a student who had learned a few words of Mayan while on holiday decided to test Murray's claim that he knew that language by uttering a sentence for Murray to translate. Murray reproached him. The student's sentence was in Lower Mayan. The language Murray said he knew was Upper Mayan.

Feynman and Murray were, at least off-and-on, friends. It was to be with Feynman that Murray chose Caltech over other universities that had made him offers. And it was Feynman who, in the late 1960s, provided some key theoretical evidence regarding Murray's quarks, supposedly inside every neutron and proton but never seen in isolation.

At the time, it was a major controversy in physics: If we can't isolate an individual quark, what sense does it make to say that individual quarks really exist? Aren't these particles within particles merely a convenient mathematical construct? These questions are part of a greater philosophical issue: To what extent are the results of experiments in modern accelerators direct observations, and to what extent are they merely agreed-upon interpretation of numerical data? After all, even ordinary particles such as electrons and protons are thought of as having been "observed" even though we "see" them only through such indirect evidence as the tracings of their path on film, or the clicks of a Geiger counter. And for more exotic particles the evidence is even less direct: They are inferred to exist from statistical blips on charts of data pertaining to the scattering of other particles. Might not a civilization on Mars, making the same experimental observations, have a completely different concept of the "reality" that underlies them? One school of philosophy, called positivism, avoids these issues by holding that only what we can sense directly can be accepted as reality. Modern physics has ventured far beyond the positivist point of view. But for many, the idea that an unobservable particle like a quark was real was pushing the envelope a little too far. When confronted with such issues, Feynman would say he had doctor's orders not to discuss metaphysics. Yet it was he who, in the late sixties, published work showing how certain experimental observations of proton behavior could be explained by assuming that protons had an internal structure of unseen subparticles—the kind of indirect "observation" of quarks that most physicists accepted as proof of their existence. Ironically, Feynman, ever the cynic, begged to disagree. Quarks had many special properties that were not relevant to the physical process he had investigated. Thus one could not conclude from his calculations that the unseen particles of his theory had those properties, that is, they were actually quarks. It could have been that Murray's theory was wrong, and that other, yet to be characterized, unseen particles existed within the proton. Because of this, Feynman refused to dub the internal particles of his theory quarks, and called them, instead "partons." This annoyed Murray, partly because of the refusal to endorse his work, and partly because the word parton is a mixture of Latin and Greek roots. But that was Feynman: fastidious about describing nature, cavalier about the rules for mixing Latin and Greek.

Though Feynman scorned the study of philosophy, it was such differences in philosophy that underlay the friction between the two men. Feynman used to say there were two kinds of physicists, the Babylonians and the Greeks. He was referring to the opposing philosophies of those ancient civilizations. The Babylonians made western civilization's first great strides in understanding numbers and equations, and in geometry. Yet it was the later Greeks—in particular, Thales, Pythagoras, and Euclid—whom we credit with inventing mathematics. This is because Babylonians cared only whether or not a method of calculation worked—that is, adequately described a real physical situation—and not whether it was exact, or fit into any greater logical system. Thales and his Greek followers, on the other hand, invented the idea of theorem and proof—and required that for a statement to be considered true, it had to be an exact logical consequence of a system of explicitly stated axioms or assumptions. To put it simply, the Babylonians focused on the phenomena, the Greeks on the underlying order.

Both approaches can be powerful. The Greek approach brings with it the full force of the logical machinery of mathematics. Physicists of this ilk are often guided by the mathematical beauty of their developing theories. And it has led to many beautiful applications of mathematics—such as Murray's classification of particles. The Babylonian approach allows a certain freedom of imagination, and allows you to follow your instinct or intuition, your "gut feeling" about nature, without worrying about rigor and justification. This aesthetic has also led to great triumphs—triumphs of intuition and "physical reasoning," that is, reasoning based principally on the observation and interpretation of physical processes, and not driven by mathematics. In fact, physicists employing this kind of thinking sometimes violate the formal rules of mathematics, or even invent strange new (and unproved) math of their own based on their understanding of experimental data. In some cases this has left mathematicians bringing up the rear—either justifying the physicists' novel use of their ideas, or investigating why their "unwarranted" use gives pretty accurate answers anyway.

Feynman considered himself a Babylonian. He relied on his understanding of nature to guide him wherever it might lead. Murray was more the Greek type—wanting to categorize nature, to impose an efficient mathematical order on the underlying data.

Though it angered Murray for Feynman to refuse to identify the internal elements of protons as quarks, this is exactly what you would expect from a Babylonian-style thinker. Feynman had explained some data by pointing out that it seemed as if an internal structure were present. He didn't see in that data any compelling reason to take the further step of identifying the internal structure as the one proposed by Murray. To a Greek-style thinker, the fact that this identification would tie in to a beautiful mathematical classification scheme was a compelling reason to make it.

Despite Feynman's characterization of these approaches as Babylonian and Greek, a similar philosophical tension has played out with many other characters and movements throughout history, for instance, among the Greeks themselves: Plato and Aristotle. Plato believed that, underlying the varied phenomena of the material world, there are eternal and immutable patterns. It is the description of these, in mathematical terms, that physicists such as Murray seek. Aristotle felt that Plato got it backward. To him, the ideal—that is, abstract—description of nature was a myth, or perhaps a convenience, and what we really ought to be concerned with were the phenomena we perceive with our senses. Like Feynman, he worshiped nature itself, not the (possible) underlying abstraction.

Feynman's distinction seemed to me to also mirror Sperry's theory about the two hemispheres of the brain. The left, seeking order and organization, is Murray, the Greek, the Platonic, and the right, perceiving patterns and emphasizing intuition, is Feynman, the Babylonian, the Aristotelian. In light of its physical root within the brain, it is no wonder that their difference in approach extended beyond physics, to the way they lived their lives. It was a life choice with which I, too, without realizing it, would soon be confronted.

In many ways Feynman was Murray's intellectual

nemesis. Though in 1981 Feynman had not yet been discovered by the popular media, in the physics world his persona had already outshone Murray for decades. The Feynman legend began when, in 1949, at age thirty, he wrote that series of papers for the *Physical Review.* Ever since Isaac Newton, you created a theory in physics by writing down an equation, or set of equations, called differential equations. Then you calculated the consequences of the theory by solving the differential equations. Quantum theories were no different. For instance, to discover what quantum electrodynamics—the quantum theory of electrically charged particles—predicted for the future behavior of an electron, a physicist in the 1940s would first describe its current, or "initial," state. This mathematical function contains information describing quantities such as the electron's momentum and energy at the beginning of a process or experiment. The goal of the theorist would be to describe these same quantities at the end of the process or experiment (that is, to calculate what is called its "final state") or at least to calculate the probability that it reaches a particular final state of interest. To accomplish this, the physicist would solve a differential equation. Feynman's formulation of quantum theory did away with the need to solve the differential equation.

In Feynman's approach, to find the probability that an electron that started in a given initial state would end up in some particular final state you add, using certain rules, contributions from all the possible paths, or histories, of the electron that could take it from the initial state to the final one. To Feynman, this was what distinguished the quantum world from the everyday, or classical, world. In classical theories a particle followed a definite path, just as objects seem to in our everyday world. The strange quantum world arises because you have to take into account extra paths. For large objects, the way you add up the paths makes only one of these important, the familiar classical path, so you don't notice any quantum effects. But for subatomic particles, such as the electron, you cannot ignore paths in which the electron travels to the far reaches of the universe, or zigzags back and forth in time. The quantum electron shoots around the universe in a cosmic dance, from present to future to past, from here to everywhere in the universe, and back. In following these paths, it ignores the orthodox rules of motion and acts as if nature had let go of the controls. As Feynman put it, even "the temporal order of events . . . is irrelevant." Yet somehow, like the music of instruments in harmony, all these paths, added together, add up to the final quantum state that the experimenter observes.

Feynman's method was radical, and at first glance, absurd. In our science-oriented culture, we expect order. We have developed a strong idea of time and space, and that time progresses from past to present to future. But underlying this order, according to Feynman, are processes that are free from following such rules. Feynman as usual would never discuss such metaphysical aspects of his theory. Later, when I got to know him, I felt I understood how he could conjure up such a theory: He himself behaved much like the electron.

Feynman's approach was hard for physicists at the time to grasp and accept. The so-called "path integrals" he had invented to sum the paths were mathematically unproven, and, at times, ill-defined. And his pictorial technique for generating answers from his theory—today called Feynman diagrams—was unlike anything physicists had seen before. Physicists demanded proof. They wanted a mathematical derivation of his formulae starting from the usual formulation of quantum theory. But he had developed his method employing intuition and physical reasoning—plus plenty of trial and error. He couldn't prove it. When he presented his method at a conference in 1948, he was roundly attacked by star physicists like Niels Bohr, Edward Teller, and Paul Dirac. They demanded the Greek approach, and here he was, a Babylonian. Yet in the end they could not ignore him: He could do theoretical calculations in a half hour that took them months.

Eventually another young physicist, Freeman Dyson, showed how Feynman's approach was related to the usual one, and it slowly caught on. Some, such as Murray himself, speculate whether Feynman's method, his path integrals and Feynman diagrams, rather than Newton's approach employing differential equations, isn't the true foundation of all physical theory.

Though to physicists Feynman was legendary, and Murray all too human, in some ways Murray had been more influential in guiding the direction of the field. This is because Murray, ever seeking order and control, had always sought the leadership role. Feynman avoided it, preferring to let his work speak for itself.

How did I fit in?

The source of my own success was my Ph.D. dissertation and several papers I wrote with a Berkeley postdoctoral fellow from Greece named Nikos Papanicolaou. Like Feynman, Nick and I explored a way of connecting the quantum and classical worlds: We discovered that the quantum world would look similar to our classical world if only we lived in a universe with many more than the three dimensions of space we are familiar with. Then we showed how certain problems in atomic physics would be easily solvable if the world had an infinite number of dimensions. And finally, we showed how to compensate for the false assumption of infinite dimensions, and find answers that are accurate and relevant to our 3-D world. When the smoke cleared, I had been stunned at the accuracy of our approach. But most of all, I was proud of our originality.

Our work had been cited a year or so earlier in an article in the semitechnical professional journal *Physics Today* by a young Princeton professor named Edward Witten, who, in the coming decade, would take the late Professor Feynman's place as the number one Yoda of the physics world (and eventually occupy Murray's old office). After that article, others began to cite our work. The number of citations grew to dozens. I lost track when it reached one hundred. I also found myself being treated with a new respect. My Ph.D. advisor was suddenly interested in the minute details of my work. Out of the blue, an old professor from my undergraduate days sent his regards. Professors started treating me as if what I said about things might be worth listening to. As the time came to think about what I would do next, the bad thoughts started coming. The doubts. Could I ever repeat my success? And then came the job offer from Caltech.

Whether Greek, Babylonian, or just native Chicagoan, I knew I had to discover my own style and approach to physics—and to life. Yet first I had to get over my feelings that my discovery was a fluke, and my success some kind of hoax, or a lucky break that would never happen again. I spent weeks in that state of mind, staring for long stretches at one journal or another, hardly turning the pages, nothing sinking in. I would go to seminars unable to focus on the topic. I would have conversations with fellow postdocs in the corridors, but barely be able to follow the simplest lines of thought.

At home I started spending evenings with a couple of neighbors who had found their niche in the world smoking joints. Edward, a thin, short Caltech physics grad, smoked away his boredom and moral qualms with a job doing weapons research, and Ramon—whom everyone called Ray—a garbage man, smoked to forget the smells he had been subjected to earlier in the day. I sat beside them, a twenty-seven-year-old has-been nervous about keeping the secret that he was, in reality, a never-was. Together, we'd watch reruns of *Columbo,* or *The Rockford Files,* secure in the knowledge that whether we paid attention or not, the bumbling detectives would always get their man.

Meanwhile, winter came, and with it the new semester and the new year. By now I would see Feynman, back from his surgery, coming and going from his office down the hall. I figured if anyone could help me emerge from my creative drought, it would be my idol Feynman. His writings had first excited me about physics, and now fate had dropped me into the department just a few doors down from him. All I had to do was walk a few steps and knock on his door. Fortunately, with all my naïveté and self-doubt, I did have pluck, or chutzpah as my parents would say. Not even living legends were unapproachable. So Feynman, who despised psychology even more than philosophy, would soon become my leading advisor on both the philosophy and the mind of the scientist.

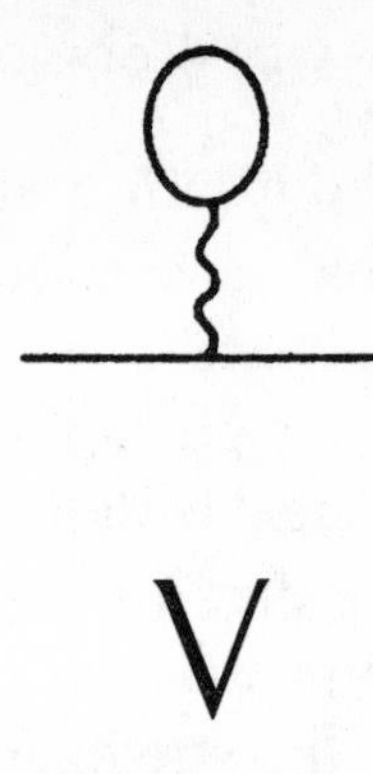

V

WHEN I FIRST SET eyes upon him, the image did not match the legend. Feynman was sixty-three—about ten years older than Murray—but he looked gaunt and aged. His long gray hair was thinning; his step lacking in energy. With my state of mind at the time I might have looked a bit like him, but Feynman's malaise was unlike my own. It was common knowledge by then that Feynman was terminally ill. In his recent surgery he had had a widespread tumor entangling his intestines removed in a marathon fourteen-hour procedure. It had been his second cancer surgery.

I stepped over to his office, knocked, and introduced myself. He was polite, and welcomed me. I had had no direct experience with death. It was hard for me not to feel pity, as I might for a deformed person I saw on the street. The thought of actually talking to a dying person made me uncomfortable. Strangely, I would find that being one did not seem to have the same effect on him. I could see right away that there was still an energy about him, a gleam in his eye. He may have had terminal cancer, but his spirit still zigzagged around the universe.

Though my heart was pounding, I was surprised at the impression he made. He didn't have that distancing sheen of brilliance that Murray had; in fact, there was nothing about him that indicated greatness. If I had run into him on the street, and hadn't seen pictures, I might have thought he was a retired cab driver from Brooklyn. I had the impression that in his younger days he must have possessed a certain earthy sexuality. After we had exchanged a few words, he mumbled a "see-you-around" and looked back down at his work. I left.

A few days later I bumped into Feynman outside the Lauritsen Lab.

"Mlodinow, right?" I was flattered that he remembered, and happy he didn't pronounce my name in some weird Russian way. I asked where he was going.

"To the cafeteria."

"The cafeteria or the Athenaeum?" I asked. Unlike the elegant Athenaeum, a place favored by Murray—and most faculty—where the men often wore suits and the servers were students, the cafeteria back then was an unremarkable joint with food I imagined you might find in an army mess hall. It was usually referred to by its more descriptive nickname, "the Greasy." Feynman gave me a look. Apparently, the Athenaeum wasn't his style. He invited me to join him at the Greasy.

The Caltech cafeteria in those days had a novel way of cooking their hamburgers. They would partially cook dozens of them around ten in the morning and leave them stacked at the back of the grill. When you ordered a burger, they would flip it off one of the stacks and more or less finish cooking it. As it turned out, this culinary technique meant that the kitchen had much in common with the microbiology lab, except that their hamburger was probably cheaper than the sterile agar used in the labs. We came in around two, near closing, by which time the burgers had been kept half-cooked and tepid for several hours. Still naïve in the ways of Caltech, I ordered two burgers, one with fries, the other with onion rings. For me, it was breakfast.

We sat down. Feynman usually drew a crowd at the Greasy, but this late there wasn't anyone else around. We sat in silence for a moment. I tried to think of something intelligent to say to break the ice. My mind was a blank. The feeling was a lot like one I'd have again many years later, accepting a computer game award in Cannes. Then, I was onstage, in a spotlight in front of thousands. I had uttered a few lines that I had prepared, and then made ready to walk offstage. But the beautiful French TV celebrity who acted as host surprised me with a question. I couldn't think of anything to say to her, not even my name. It was as if the spotlight had saturated my neural circuits, making intelligent thought impossible. I wished I were pretty enough to charm everyone with my smile, then wave and walk off like a star. Instead, I just stood there embarrassed as she finally answered her own question.

With Feynman I got off easy. He looked at my tray. Then he looked at me and smiled.

"I used to overeat," he said. "If I really liked the food I'd eat so much I would feel uncomfortable. That was dumb. I don't do it anymore."

"I think I can learn a lot from you," I said, then realized how stupid it must have sounded.

"Yeah, well I don't know what's good for anybody except myself."

More silence. My mind raced. I knew that before long, others would join us, and my chance to get his advice would be gone. I wanted to ask, "How do I know if I'm smart enough to be here?"

Instead I said, "Read any good books lately?"

He just shrugged.

"I've been reading about the process of discovery," I told him, trying to keep the conversation alive. I was in the midst of Arthur Koestler's *The Act of Creation*.

"Learn anything?" he asked. He was interested. That was Feynman, always interested.

"I'm having some trouble getting my research on track, and so I thought it might help."

"Yes, but did you *learn* anything?"

He was mildly annoyed now, because I hadn't answered his question. I felt rebuffed. I wasn't yet sure what I had learned, so I told him about the passage I had just finished. I tried to make it sound dramatic.

"It took place in Berlin, 1914. Imagine a cold spring morning. Outside church bells chime. In his office at Berlin University, Einstein ponders the still-unfinished theory of relativity. In a lab not far away, in a tall steel cage, a young chimpanzee named Nueva pushes banana skins together in a heap with a stick. In a few years, this episode will be retold in a famous book, *The Mentality of Apes.* But, as she glances around the room, Nueva doesn't care about fame. Her world is simple. Eat, drink, sleep . . ."

"Don't forget sex," he added with enthusiasm. I found that Feynman often found ways to interject the subject of sex. I was glad my story was holding his interest.

"Yes, and have sex, find companionship. But right now she is hungry, and banana skins won't do. As Nueva studies her plight, a professor named Koehler studies her. He, like Nueva—and Einstein—has a hunger to satisfy, and his notes are destined to feed many books and papers. Koehler offers bananas to Nueva, only he doesn't do her the favor of placing the food inside her cage. Instead, he places it on the floor outside, beyond her reach."

"A cruel fellow," said Feynman.

"He's challenging her," I said. "To eat, Nueva will have to discover how to get the bananas. First, she does the obvious. She steps to the bars and reaches out. She strains her arms and grasps at the fruit, but the bananas are just out of reach. She throws herself to the ground and rolls on her back in despair. Not far away, Einstein is nine years into his work on the theory of relativity, and still two years from his great breakthrough."

"And probably feeling a lot like Nueva," said Feynman.

I nodded and smiled. Here we were, Feynman and I, conversing about the frustrations of research. Me and Feynman, peer-to-peer! We were connecting. I was happy.

I continued, "Seven minutes pass. Nueva suddenly stares at the stick. She stops moaning and grabs it. Nueva thrusts it out of the cage, just beyond the fruit, and pulls it to within arm's reach. She has made a discovery."

"And what did this incident teach you?" Feynman asked, not letting me off the hook. I was consciously pleased when I realized that intelligent thoughts were now actually forming in my head in response to his question.

"Nueva had two skills. One was pushing things around with a stick. The other was reaching out through the bars for things. Her discovery was that she could put the two disparate skills together. It turned her old tool, the stick, into an altogether different kind of tool. Just like Galileo did when he used the telescope, which had been invented as a toy, to look at the sky. A lot of discoveries are like that, new ways of looking at old things, or old concepts. But the raw materials for the discovery had always been there, which is why the discoveries may seem startling at the time, but are simple and obvious to later generations. So I guess I learned something about the psychology of discovery. Something I might hope to apply."

He looked at me for a moment.

"You're wasting your time," he said. "You don't learn how to discover things by reading books on it. And psychology is a bunch of bullshit."

I felt as if he had slapped me. But then, after a

pause, he looked me in the eye, and said gently and with a sly grin, "What I would learn from your story is that if an ape can make a discovery, so can you."

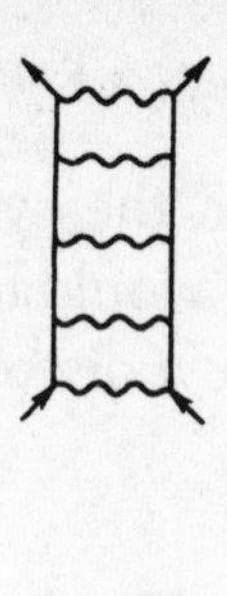

VI

MANY WEEKS PASSED, and I became friendly with Feynman, but I did not become his friend. We began to talk more easily, mainly because I became less nervous around him. I had asked if I could tape some of our conversations because I wanted to write something about him. I didn't know what—a magazine article maybe. I wasn't sure I could do physics anymore, but I had always loved to write. It was an escape for me, like going to the movies. And he didn't seem to mind. He always liked an audience.

It was a cool day. The campus was quiet; the few students walking weren't talking. Inside, Feynman's office was utilitarian. The blackboards were covered with mathematics—mostly with Feynman diagrams like those he had invented in his youth. There was a desk, a couch, a coffee table, a couple of bookshelves. Nothing seemed opulent. Nothing indicated that he was one of the most famous and respected scientists of the twentieth century. He was speaking to the point that had bothered me most—did I have that something special that it takes to be a scientist?

Feynman said:

Don't think it is so different, being a scientist. The average person is not so far away from a scientist. He may be far away from an artist or poet or something, but I doubt that too. I think in the normal common sense of everyday life that there is a lot of the kind of thinking that scientists do. Everyone puts together in ordinary life certain things to come to conclusions about the ordinary world. They make things that weren't there, such as drawings, such as writing, such as scientific theories. Is there something common in the process? I don't see such a big difference between that and the scientist's work.

For instance, any ordinary person can lie, and lying takes a certain imagination. And you have to make up a story that is sort of reasonable with nature, and it might even have to fit with certain facts. Sometimes they do a good job. They don't have to be scientists or writers.

Is science anything more wonderful than the person who says, "Mary hasn't come home yet, I betcha she went to the Loaf and Ladle for lunch, because she likes to go there? Let's call there." You call there and Mary is there. Is that creativity? The average person puts together ideas of their experience to see something else, or some relationship, and suddenly notices that the twitch that little Mary has always comes when she is talking about school. Then they do something with that realization. All ordinary life and behavior involves human activity which seems to me to be very similar.

Scientists do think in a constructive way. You ask a scientist some question and it worries him. He doesn't worry in the sense that a normal person sometimes worries, like "I wonder if this sick person is going to get better." That's not thinking, that is just worrying. The scientist tries to build something up. Not just to worry about something, but to think something out.

The scientist analyzes something like a detective does. Like a detective trying to find out what happened when he wasn't there, given clues. We are trying to figure out what nature is like from clues given by experiments. We have the clues and we try to figure it out. It is more analogous to detective work than anything else.

Somehow I did not picture Feynman as Sherlock Holmes. That was more like Murray, a person who always seemed to walk around muttering, "Elementary . . ." to whomever he was with. Murray was from the I-can-do-it-because-I-am-smarter-than-everyone-else school of physics. Of course, Murray *was* smarter than everyone else. But I wasn't. Feynman dressed and spoke like more of a blue-collar, regular-guy physicist, which was more my type. With that thought the detective metaphor suddenly made sense to me—and I found it encouraging. I knew that there were fumbling detectives like Rockford and Colombo—or regular guys like Sam Spade—who nevertheless managed to reveal the mysteries of the world around them.

Still, when I got back to the apartment that night I

suggested to Edward and Ray that we go to the library to rent a Sherlock Holmes movie, figuring he was a better role model for a physicist than Rockford. It was the days before VCRs, so we would actually check out film and a projector and project the movie on a wall on the outside of the building. From that week on, my neighbors and I moved outside every Friday night and watched the same flick—*The Hound of the Baskervilles*. With joints and beers, we would sit under the palm trees beside the pool and revel in its shadowy, black-and-white frames. Edward would occasionally dress as Sherlock, though the substance in his pipe was not conducive to Holmes's kind of tight, logical analysis. Together we'd call out otherworldly Basil Rathbone's camp lines in advance, like the audience in a 1939-vintage version of *The Rocky Horror Picture Show*. By the end of the evening, I would be lost in a world halfway between Pasadena decadence and Old World decorum, and marvel at the power of film.

Feynman continued:

> Really all we do is a hell of a lot more of one particular kind of thing that is normal and ordinary! People do have imagination, they just don't work on it as long. Creativity is done by everybody, it's just that scientists do more of it. What isn't ordinary is to do it so intensively that all this experience is piled up for all these years on the same limited subject.
>
> A scientist's work is normal activities of humans carried out to a fault, in a very exaggerated form. Ordinary people don't do it as often, or, as I do, think about the same problem every day. Only idiots like me do that! Or Darwin, or somebody who worries

> about the same question. "Where do the animals come from?" Or, "What is the relation of species?" A scientist works on it, and thinks about it for years! What I do, is something that common people often do, but so much more that it looks crazy! But it's trying to find the potentiality as a human being.
>
> For example, neither you nor I have muscles that stand way out on our arms like these fabulous guys. For us that would be impossible. Well they work and they work and they work on it. In that case, it might be a fault. How big can you make those muscles? How can you make the chest look great? They try to find out how far you can go. And therefore, they do something with an intensity that is out of the ordinary. It doesn't mean that we never lift weights. All they do is lift weights more. But, like us, they're trying to find the greatest potentiality of human beings' activity in a certain direction.

The scientist as a brain jock? Did I believe him? Is creative genius a form of synaptic sweat?

I had gone into physics and through my studies thinking physicists were something like mystics. After all, the physicist's pen can shake theology with a new view of creation, or change the world with an invention, like the radio, the transistor, the laser . . . or the Bomb. The physics lore you get in school encourages this view: We read about Einstein, his IQ off the charts, employing pure logic to derive the connection between space and time; we heard Niels Bohr described, because of his physical intuition, as having a direct line to God; we toasted Werner Heisenberg, who formulated the uncertainty principle that shook the foundations of mechanistic philosophy. Amongst my friends, these physicists were all mythic heroes.

People picture scientists in white coats. Physicists, at least, don't wear them, but in a way, I subscribed to the same basic misconception: that scientists were somehow different from other people. I read about their theories in the tight logical development that comes only long after the fact. I knew nothing of their insecurities, their false starts, their confusion, their days in bed with a bellyache. Even as a graduate student I never got to know any faculty members as people. They were there to ask questions of, but always separated by the same partition that separates the rich and the poor. Now I was on the faculty myself, a real scientist, and that is what seemed so strange. I didn't see myself as different, so, if scientists were different, how could I be a scientist? Feynman was saying don't worry, they aren't. It was a simple realization, and a comfort.

There is a flip side to the comforting knowledge that everyone is just stumbling through the fog, and that is that it is a pretty good bet that many of them are not stumbling in the right direction. Who is going down the blind alley and who is on the road to success? Whose work will be remembered and whose forgotten? What is worth doing, and how do you know? I didn't have the answers, but I thought back on the pep talk the division chair had given me. Explore, he said. Check out what other people are doing. I decided to open myself to others.

VII

THE FIRST GUY I tried to connect with was a fellow named Stephen Wolfram with a position similar to mine. We met for lunch at a place that called itself an Italian deli. Wolfram ordered a plate of rare roast beef. They gave him maybe a pound of it. No bread. No potato chips. No pickle. Just a pound of red meat. I had a regular sandwich, potato chips and all. Despite our divergent tastes in food, we got into a pretty friendly conversation. He seemed like a nice enough fellow at first, but when we got to talking I discovered some things that alarmed me. Namely, he had studied at Oxford, published his first scientific paper at the age of fifteen, and received his Ph.D. in theoretical physics from Caltech by the age of twenty. No, I decided. We could never be friends. Years later, I would read about him often, founding a wildly successful software company, and then publishing a famous book, an outgrowth of his pet theory, cellular automata. Ordinary people? I wondered if Feynman had met this guy.

A few days later I got to my office with a headache. I had been up till four with Ray, who was depressed because he couldn't find a girlfriend. Lately he seemed extraordinarily devoted to this task. He'd mutter to himself, sometimes in Spanish, the only reminder that his given name was Ramon and not Ray. If a love song came on the radio he would yell curses, or change the station, or, once, smash the radio. He thought about his woman problem day and night. It consumed him. I thought about him, employing Feynman's analysis, as a scientist. His field was love, and, like Darwin or Feynman, he worried about the same question all the time, in his case the question of finding a mate. Ray had been talking suicide, and, since he owned a gun, I thought it my duty to make sure he didn't use it. So I kept him off drugs, and we drank martinis instead. We found that we could commiserate in that we had similar problems plaguing us. Neither of us could find the mistress we desired, the mistress in my case being a good problem to work on.

At my office, it didn't help my headache that I could hear Murray through the wall screaming at someone over the phone. It appeared to be someone at the bank, a clerk of some kind, and he or she was being dense about something. It could really bother Murray when people didn't know things, or grasp ideas as fast as he did. Unless of course it was Feynman, in which case Murray reveled in it. And since Murray had encyclopedic worldly knowledge, and Feynman's factual knowledge was focused on math and the sciences, there was plenty Murray could talk about where Feynman would be at a disadvantage.

I chewed a few aspirin and wondered what I was going to do. I had had periods between papers before, times I would spend just reading and thinking, trying to come up with a good idea, or a good problem to try to solve. That is normal for a theoretical physicist. The inability to concentrate isn't. I decided to pay a visit to a young professor down the hall. Maybe, I thought, we could collaborate on something. He seemed accessible, and had produced a famous Ph.D. dissertation having to do with the strong force.

One of the attractions of physics is the magnitude of the ideas you ponder. It may seem like a yawner to spend your day fiddling with mathematical expressions, but it becomes exciting when you realize that in studying the strong force you are exploring a power as great as any you'll find in even the most speculative science fiction. Without the strong force, the electric repulsion between the positively charged protons in the nucleus would blow apart every atom in the universe, except those of hydrogen gas, whose nuclei consist of lone protons. When you think of it that way, the power and potential of what you may discover seems almost unlimited.

It is the strong force binding quarks to each other that physicists believed was the reason that Murray's quarks were never seen in isolation. But there was a problem with this explanation: According to the experimental observations, when particles like protons were smashed into each other, they behaved as if the particles inside them—what Feynman had called partons, but which everyone else believed were quarks—could jiggle around freely. How could they move freely if they were bound together so tightly? Because it was so difficult to calculate the consequences of quantum chromodynamics—the theory of the strong force—the answer to this question was not at all apparent. This young professor down the hall had done groundbreaking work on this problem. The answer, it turned out, was that according to quantum chromodynamics, the strong force, unlike the other fundamental forces, grows stronger with distance. If you could pull two quarks an inch apart (which you can't), they would experience an unimaginably strong attraction; two quarks within a proton hardly affect each other, and behave as if they were free.

To escape the effects of the strong force, you don't run away, you move closer. Though a novelty in physics, this behavior was much like the human forces that acted upon me at Caltech. I was there, supposedly free to do whatever I wanted. And as long as I acted like a serious scientist working on important research I did feel free. Yet I did not feel free to say something stupid. I did not feel free to fail. I did not feel free to be anything except obsessed with research—and not just any research.

In the culture of physics I had grown up with, there was a hierarchy of respectability. My office was on the floor that housed the elementary particle theorists—those who, like Feynman and Murray, work on the theory of the fundamental forces and particles in nature. They tend to look down on others, like biologists, or chemists, or most other physicists, who are applying, rather than discovering, fundamental laws. In this view, even solid-state physics, which has led to discoveries such as the transistor, and thus our modern digital age, is dismissed as a less worthy calling. "Squalid state physics," Murray called it.

I imagined you could chart this cultural landscape along the lines of Saul Steinberg's classic *New Yorker* magazine cover looking westward from Manhattan. I pictured, in the foreground, at the center of the world, the different aspects of elementary particle theory—like the buildings of Manhattan. This is what Feynman and Murray and most of those on the floor worked on. The surrounding areas—off in New Jersey somewhere—represented mathematics and other areas of theoretical physics. In the vast and distant middle of the country were the marginalized great plains of experimental physics. Finally, on the far coast, were some tiny structures—applied physics, the life sciences, and other professions hardly worthy of attention. As long as I stayed near the world's center, I was free to move about. But the farther my research might stray from it, the stronger the force I would feel pulling me back.

Feynman always made a point of ignoring such forces. He was interested in all of physics, in other sciences, and in many other creative endeavors. Even socially, he would not conform. Expected to behave with a certain professorial decorum, he would go to a strip club to work on his physics. At the strip club, he'd be expected to drink alcohol, or maybe cavort with the strippers. But he didn't drink, and was faithful to his wife. What I didn't realize at the time was that I, too, had the power to ignore the force of other people's expectations.

I didn't have the insight back then to apply this analysis of the strong force to myself, yet this young professor's idea appealed to me. I also figured that since he had had early success, like me, and he had succeeded in making it to the next step—as a tenured professor at Caltech—he would be a promising mentor.

I stepped into his office. Several houseplants and a poster of the Huntington Gardens—a famous nearby botanical garden—adorned the room. It was only the second time I had ever seen plants in a physicist's office. The other was a mathematical physicist I once knew but he doesn't really count because his plants were all dead from lack of water.

The young professor was a large, rotund fellow. He looked cheerful. After a bit of small talk, I asked what he was working on these days, trying to be as nonchalant as possible. Most researchers are happy to find collaborators, but no one wants a desperate collaborator. My nonchalance must have been exaggerated though, because he gave me a funny look.

"I'm just going around," I said, "getting acquainted with what everyone on the floor is doing."

"I get it." He smiled. But he still didn't answer.

"So . . . what are you working on?" I asked once more.

"Oh, you won't want to work on that."

"You never know," I said.

He kept smiling, but he didn't speak. I stared at him as a driver might stare at a streetlight, waiting for it to turn green. But the light didn't change.

I read once about a study that concluded that the trait most correlated with success in graduate school is persistence. I felt that in sociology researchers often had an excess of this trait themselves—they persisted in drawing conclusions beyond the point of statistical validity. Still, being a persistent fellow, I often took comfort from that study.

"So what do you work on?" I persisted.

He shrugged. "Oh, these days . . . mostly gardening." His smile remained undiminished through his answer.

Out in the corridor I supposed he earned his keep teaching, but I looked down upon him. To teach science was not to be a scientist, and to me, back then, not worthy of his position. From then on, I always thought of him as Professor Gardening.

I ran into my friend Constantine. He was a postdoctoral fellow from Athens. His father was Greek, but his mother was Italian, and he seemed to have inherited from her an impeccable sense of style, both in the way he dressed and in his approach to physics.

"Don't you know about him?" he whispered. "He's burned out. They gave him tenure right after graduate school because everyone was fighting to get him. Turns out he's just a one-trick pony." Constantine smirked.

A one-trick pony. I smirked back out of obligation, but I was thinking, *just like me.* Except that nobody made the awful mistake of giving me tenure. In a few years I would be completely lost, I imagined, and have to take a depressing job like my neighbor in the defense industry. I couldn't see myself designing missiles, though, at least not without final say on against whom they were used.

My head still ached, so I went to Helen, the department secretary, looking for more aspirin. She had the office on the other side of Murray's from mine—the one between Murray and Feynman—and had been in the department about as long as they had. As I approached her office I heard her saying to someone inside, "You sure gave that bank teller a hard time."

And then Murray's voice, "Oh, you heard that?"

And Helen: "How could I not?"

Murray emerged from her office. He nodded. I nodded. I went in to see Helen.

"You have a headache?" she said when I asked for the pills. "I'm not surprised."

I gave her a look: What did that mean?

"If you don't mind my saying so, you haven't looked very happy lately."

"Oh, I'm just . . . struggling with what I'm going to work on next."

"Well I don't know anything about physics, but it seems to me that everyone does that. At least the ones who haven't given up."

I said, "I bet Feynman doesn't."

"Professor Feynman? Why, he's had long dry

spells. Everyone knows that—at least everyone here. But he always bounces back. I'm sure you will, too."

She gave me the pills. Then she said, "Or if not, you'll find something else to do with your life. You're still young."

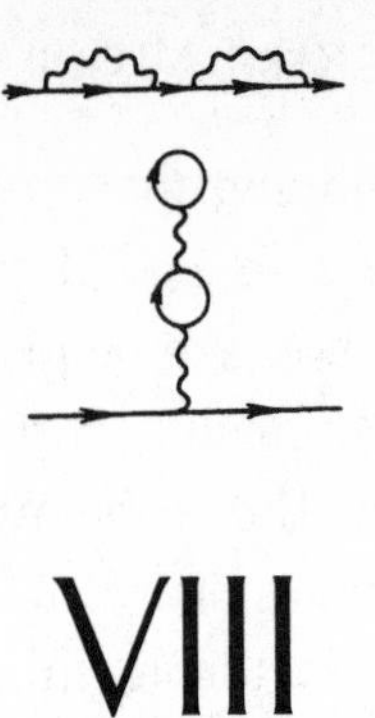

VIII

IN HIS YEARS OF practicing physics, Feynman had solved several of the toughest problems of the post-war era. In between, I confirmed, there were some prolonged periods of inactivity. And indeed, he always bounced back. And whereas Murray worked almost exclusively in the field of elementary particle physics, Feynman had made important contributions in many areas—low-temperature physics, optics, electrodynamics. He seemed to have a knack for finding the right problem to work on, and at the right time. I wondered, what was his approach? What took more talent, choosing the right problem, the issue I was now struggling with, or finding the solution? And once he settled on a problem, what did it take to solve it?

> When you first came here and asked to discuss how I approached a problem, I panicked. Because I really don't know. I think it's like asking a centipede which leg comes after which. I have to think a while, try to look back and quote some problems.

In some cases finding the problem you work on could be a result of a very good creative imagination. And solving it may not take nearly the same skill. But there are problems in math and physics where there is the reverse situation. The problems become sort of obvious and the solution is hard. It's hard not to notice the problem and yet the techniques and methods known at the time and the amount of information known to people is a small amount. In that case the solution is the ingenious thing.

A very good example is Einstein's theory of relativity and gravity—the general theory of relativity. With relativity it was clear that they had to combine somehow this theory of special relativity, that light travels at a certain speed, c, with the phenomenon of gravitation. You can't have that—you can't have the old, Newtonian gravity with infinite speeds, and the relativity theory that limits speeds. So you have to modify the theory of gravity somehow.

Gravity had to be modified to fit the theory of relativity, that light undergoes motion at a certain speed. Well, that's not much to start with. How to do it?! That was the challenge!

That this had to be done was obvious to Einstein. It wasn't obvious to everyone, because to them the special theory of relativity wasn't yet obvious. But Einstein had gotten past that. So he saw this other problem. It was obvious, but the way of solving it, that took the utmost imagination. The principles that he had to develop! He used the fact that things were weightless when they fell. It took a very, very lot of imagination.

Or let's take the problem I'm working on now. It's perfectly obvious to everybody. We have this mathematical theory called quantum chromodynamics that is supposed to explain the properties of protons and neutrons and so on.

In the past if you had a theory and wanted to find out if it was right you just took it out, looked at what happened in the theory, and compared it to experiment. Here, the experiments are already really done. We know lots of properties of the protons. And we have the theory. The difficulty is that it's new, and we don't know how to calculate the consequences of this theory, because we haven't got the mathematical power.

To invent a way of doing it. Now how do you do that? You have to create or invent a way to do it. I don't know how to do it. Here the problem is obvious, and the solution is hard.

It took many pieces of imagination to find this theory, people noticing patterns and gradually discovering things, ultimately the quarks, and then trying to find the simplest theory. So there was a long history that produced this particular problem. It took us a long time to get here, but now our noses have been kind of rubbed in it.

Rubbing our noses in it. It was an interesting expression for him to use. I found it comforting when Feynman revealed that he, too, got frustrated.

I am working on this very hard problem now, and have been for the last few years. The first thing I tried to do with this problem is try to find some sort of mathematical way of doing it, solving some equations. How did I do that? How did I get started on figuring it out? It's probably kind of determined by the difficulty of the problem. In this case, I just tried everything. It's taken two years, and I've struggled with this method and that method. Maybe that's what I do—I try as much as I can different kinds of things that don't work, and if it doesn't work I move on to some other way of trying it. But here I realized after trying everything that I couldn't do that. That none of my tricks worked.

So then I thought, well, if I understood how the thing behaved, roughly, that would tell me more or less what kind of mathematical forms I might try. So then I spent a lot of time thinking about how it worked, roughly.

There are also some psychological things there. First of all, in my later years I take only the most difficult problems. I like the most difficult problems. The problems that nobody has solved, and therefore the chances that I'm going to solve it are not too high. But I feel now that I've got a position now, the tenure, I don't worry about wasting the time it takes to work on a long project. I don't have to say I've got to get my degree in a year. It's true that I may not last so long physically, but I don't worry about that.

His illness was always there in the room with us, an angel of death patiently waiting for his time to run out.

The next psychological aspect is, I have to think that I have some kind of inside track on this problem. That is, I have some sort of talent that the other guys aren't using, or some way of looking, and they are being foolish not to notice this wonderful way to look at it. I have to think I have a little bit better chance than the other guys, for some reason. I know in my heart that it is likely that the reason is false, and likely the particular attitude I'm taking with it was thought of by others. I don't care; I fool myself into thinking I have an extra chance. That I have something to contribute. Otherwise I may as well wait for him to do it, whoever it is.

But my approach is that I'm never the exact same as someone else. I always think I have an inside track, I always try another way. And I think that because I'm trying another way that's it. They haven't got a chance. It's exaggerated. And I have to work myself up to this exaggeration. I always consider it something like Africans when they were going out to battle, to beat drums and get themselves excited. I talk to myself and convince myself that this problem is tractable by my methods and the other guys are not doing it right. The reason they haven't gotten it is that they aren't doing it right. And I'm going to do it a different way. I talk myself into this, and I get myself enthusiastic.

The reason is, when there is a hard problem, one has to work a long time and has to be persistent. In order to be persistent, you have got to be convinced that it's worthwhile working so hard, that you're going to get somewhere. And that takes a certain kind of fooling yourself.

This last problem, I really did fool myself. I haven't gotten anywhere. I couldn't say my approach is very good. My imagination is failing me. I've figured out

> qualitatively how it works, but I can't figure out quantitatively how it works. When the problem is finally solved, it will all be by imagination. Then there will be some big thing about the great way it was done. But it's simple—it will all be by imagination, and persistence.

People who have never worked in physics tend to describe it with words like dry, exact, and precise. Real-life physics is as far from that as is the practice of law from the theoretical debate in law school, or the practice of medicine from the theory of physiology and disease. The law might consist of definite rules, but its application is subject to interpretation, incomplete knowledge, practical considerations, and the psychology of those in judgment. Medical science might detail the symptoms of a disease, but few patients step into their doctor's office quoting textbook presentations of their malady. With experience doctors learn how to make judgments. Physics is also an art. Few real physics problems can be what you would, strictly speaking, call "solved." To a physicist solving a problem involves judging which aspects of a phenomenon are its essence, and which can be ignored, what part of the mathematics to be faithful to, and what to alter. For instance, a hydrogen atom consists of an electron orbiting a single proton. It is the only one of the hundred-plus types of atoms whose quantum equations can be solved exactly. And if you do something as simple as placing the hydrogen atom in a magnetic field, then the equations, altered to include the magnetic field, cannot be solved.

Take the problem of finding the light emitted by a

hydrogen atom in a magnetic field. You have to simplify. You might begin by assuming the magnetic field is what's essential, and drop the mathematical terms that involve the proton, or you might begin by assuming the effect of the proton is dominant, and drop the terms that represent the magnetic field. Or, as I did in my Ph.D. dissertation, you might rewrite the equations as if the world had infinite dimensions. To solve a physics research problem involves assumption after assumption, approximation upon approximation, and those great leaps of imagination people call thinking outside the box. It involves the ability to move forward, follow your intuition, and accept that you don't fully understand what you are doing. And most of all, it entails believing in yourself.

Feynman's approach to solving quantum chromodynamics was to write down a simplified form of the theory—and see what the properties of the theory were under that assumption. Feynman's work on the problem was reminiscent of one of his most famous earlier works—his theory of liquid helium. The problem was to explain theoretically some pretty bizarre properties of that fluid. For instance, it did not boil, and if you put it in a beaker, it would creep up the sides and spill out until the beaker was empty. After seeing physicists frustrated trying to solve this problem directly, in his usual Babylonian style, Feynman decided the best approach is to "wave our hands, use analogies with simpler systems, draw pictures, and make plausible guesses." This was Feynman's trademark: not powerful mathematics, but powerful imagination, combined with physical understanding. He solved the helium problem in a series of famous papers in the mid-1950s. He was obviously hoping to repeat that success here.

Feynman did not live to solve the problem of quantum chromodynamics. And in the twenty-plus years since our discussion, no one else has solved it, either. Today the only new results calculated from the theory do not come from a deeper understanding or a mathematical solution of the theory, but from the continued application to it of ever more powerful computers.

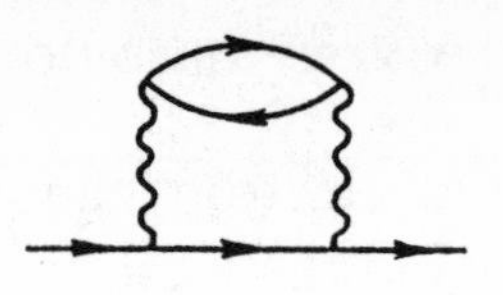

IX

As I CONTINUED my search for a problem, I thought about what Feynman had said about an inside track. What are my strengths? I was always more mathematically inclined than most of my fellow students. I was also a rebellious type—drawn by my nature to anything that went against the grain of accepted wisdom. Most of the other faculty on the floor were working, like Feynman, on discovering better ways of solving problems in quantum chromodynamics. This quest involved mostly ordinary mathematics, and was considered one of the most important problems of the day.

But there was also one professor, John Schwarz, whose research involved quite exotic mathematics, and was completely outside the mainstream.

There are four known forces in nature—electromagnetism, gravity, the strong force, and its subnuclear partner, the weak force. Physicists have a theory describing the interactions caused by each of these forces—quantum electroweak theory, a generalization of quantum electrodynamics, describes both electromagnetism and the weak force; general relativity, which is not a quantum theory, describes gravity; and quantum chromodynamics describes the strong force. The belief that all natural phenomena can be explained by fundamental physical law is called reductionism. The belief in reductionism is widespread in physics, and cuts across "party lines," from the Greeks like Murray to the Babylonians like Feynman. That means that most physicists believe that nothing happens in the universe that is not the result of one or more of the four fundamental forces—from the birth of a child to the birth of a galaxy. Given that most physicists hold this view, developing theories of the four forces is about the most important pursuit a theoretical physicist can undertake. Schwarz worked on a single theory that, if true, would subsume (and alter) all these theories. His new theory would rewrite them in one fell swoop, replacing them all with just one, all-encompassing theory.

Considering how different the four forces are, a single theory describing all of them may seem to be a far-fetched goal. For instance, the electromagnetic force can attract or repel, but the gravitational force always attracts. The strong force grows weaker at short distances, whereas the gravitational and electromagnetic forces grow stronger. And the forces also have an almost unimaginable range of strengths: The strong force is about a hundred times stronger than the electromagnetic force, which is a thousand times stronger than the weak force, which is billions of billions of billions times stronger than the gravitational force. The four forces also play different roles in our lives, and in the functioning of the universe. Gravity is, of course, what holds us to the earth, and is responsible for the ebb and flow of tides. But its most important effects in nature are on the cosmic scale. Gravity causes planets to form and to orbit their stars, and enables the nuclear furnace in a star's core that gives the light and warmth that lead to life. And long before their planets existed it was gravity's squeeze that caused these stars themselves to coalesce. The electromagnetic force is important to us mainly on the atomic scale. The electromagnetic forces amongst atoms and molecules, for instance, makes objects visible, allows oxygen to bind to red blood cells, and stops your hand from going through the wall when you lean on it. It is the force that lends to materials most of the properties they possess. And it is the harnessing of this force, mostly in the twentieth century, that accounts for most of the conveniences of modern times: from lights to telephones to radio and television to computers. The other two forces, the strong and weak forces, govern a world that exists on scales far smaller than even the atomic world of electromagnetism: the inside of the nucleus. The weak force governs the radioactive decay of the nucleus called beta decay. The strong force is responsible for atomic energy. It is the power of this force, unleashed from the nuclei that correspond to less than a third of an ounce of uranium, that destroyed the city of Hiroshima.

How could these four forces be described by a single theory? History provides a lesson here: In a way there are really five forces, but we speak of only four because the first unification of forces happened so long ago. It was the unification of the theories of electricity and magnetism, a kind of "prequel" to the present quest. The story goes something like this: Long, long ago (the sixth century B.C.), in a faraway land (ancient Greece), the simplest electromagnetic phenomena, magnetism and static electricity, were studied by a wise philosopher named Thales. From his time until the nineteenth century, human beings learned more and more about electricity and magnetism, but nothing indicated to them that these were anything other than two separate classes of phenomena. Together with gravity, electricity and magnetism constituted the three known forces of nature. Then, around the year 1820, several scientists in different parts of Europe discovered that wires carrying electric currents had mysterious magnetic properties. This was a strong hint that the forces of electricity and magnetism were related, but no one quite knew how. In the next few decades all these mortals could conjure up to describe the effects they had seen was a hodgepodge of empirical laws. Then in 1865 a Scottish physicist, barely five foot four, named James Clerk Maxwell used this hodgepodge of laws to lead him to a wondrous set of equations. In just a few lines, they showed the world how electric and magnetic forces arose from electric charges and currents—and, most important, from each other. Maxwell had thus produced a unified theory of two of the three ancient forces, electricity and magnetism, or, as we now call it, electromagnetism.

History also shows that Maxwell's unification was

more than a thing of theoretical beauty: A study of its implications revealed revolutionary new effects. For instance, his equations indicated that accelerating charges could produce waves of electromagnetic fields. These waves always moved at the same speed—which his calculations showed to be the speed of light. This provided to Einstein the inspiration for his theory of special relativity. And once Maxwell discovered that light is an electromagnetic phenomenon, it became clear that there could also exist other kinds of electromagnetic waves. This paved the way for German experimentalist Heinrich Rudolf Hertz to create the first radio waves, and eventually for the invention of such technologies as radio, television, radar, satellite communications, X-ray machines, and microwave ovens. In his *Lectures on Physics,* Feynman wrote, ". . . there can be little doubt that the most significant event of the 19th century will be judged as Maxwell's discovery of the laws of electrodynamics."

Physicists call a single theory that explains all the forces of nature a "unified field theory." It is worth taking a moment to think about what this means. For a theory to be a unified theory it has to go beyond the description of the individual forces to describe the relationship of the forces to each other, as Maxwell did when he showed how electric forces could create magnetic forces and vice versa.

Most physicists seeking a unified field theory demand even more: They seek to show how all the forces of nature arise from a single more fundamental force, or underlying principle. Though there is little experimental evidence that this is actually true of nature (or that it isn't), they seek such a theory anyway, out of an aesthetic sense, or out of faith that somewhere there is a single key to all of nature's laws. Such a unified theory would be the ultimate triumph of Greek-style physics. It is in the search for such a theory that Einstein spent most of his life, his post-relativity years, gradually drifting from the mainstream of physicists, who were more focused on more practical issues.

Beyond mathematical beauty and the potential discovery of new physical phenomena, a unified field theory also promises to answer fundamental questions about why we exist at all. It is the balance of the four forces of nature, their relative strengths and varied properties, that allows the universe to exist as we know it. For instance, suppose that the gravitational force were not so feeble compared to the strong force. Then stars would compress further and their nuclear fuel would burn out much more quickly, preventing the evolution of life. On the other hand, if gravity were much weaker, electromagnetic repulsion would prevent matter from coalescing into stars at all. If the strong force were not so much stronger than electromagnetic forces, most atomic nuclei would disintegrate. And if the numbers of electrons and protons in matter were even one percent out of balance, the electromagnetic force between you and someone a yard away would be greater than the weight of the earth. The forces of nature are disparate, but in fine balance. Why? Though separate theories can describe the individual forces, only a theory encompassing all the forces can answer this fundamental question of existence.

When Einstein started seeking a unified field theory, he was at a huge disadvantage: The strong and weak forces had not yet been discovered. But by 1981, electromagnetism and the weak force had been united in a single theory, and physicists had ideas about how to include the strong force. Progress toward a unified theory was tantalizing. Thirty years after Einstein's death, his quest gained a new popularity. The term "a theory of everything" entered physicists' vocabulary. The biggest obstacle to success, everyone agreed, was gravity. Not only didn't physicists know how to include gravity in a unified theory, but there still existed no quantum theory of gravity, even as an isolated force. Unless you believed John Schwarz. Schwarz claimed that his theory could unite all the forces, even gravity, in a single quantum theory.

The theory that was Schwarz's obsession was called string theory. The strings in string theory have little relation to ordinary strings, thin lines of fiber you might tie around your finger to remind you to buy milk on your way home. The physicist's strings were first proposed by Japanese physicist Yoichiro Nambu and American physicist Leonard Susskind in 1970. The idea was that what appeared to be a point particle might really be a tiny, undulating string. What could be the use of such a strange idea? At first, its use seemed to be that it might solve the old problem caused by the experimentalists, who kept discovering new particles. Even the number of quarks, with which

Murray was able to explain the existence of a large number of particles in terms of far fewer, had had to be greatly increased in the years since he had first proposed them. So the early allure of string theory was closely related to an idea that Murray helped originate in the 1950s, even before he came up with quarks—that all these particles may simply be alternative forms of the same thing.

In string theory there is one and only one theory that encompasses all forces, and one and only one fundamental particle—the string. Its properties would depend on the state of vibration it is in, just as the mode of vibration determines the sound created by a violin string, but in this case different states of vibration would manifest themselves as different particles, instead of different sounds. This one entity, the string, would thus account for the wide variety of particles in nature and explain the forces they react to.

From the mathematical form string theory took, there were strong indications that it held the promise of being a unified field theory of all forces, even gravity. To some, like Schwarz, this seemed to be a miracle. But these were only general properties of the theory, not predictions you could test in the lab. So the most important question remained open: Was string theory correct?

You might think that this would be an easy thing to check. You look closely at a particle. Is there a little string dancing around in there, or isn't there? But elementary particles are so small we cannot see them with enough precision to make out such finer structure. It is analogous to the reason that, from a great distance, that violin-shaped mole on your nose might look like the tiny beauty spot your mother always said it was. Still, the fact that we cannot check directly whether particles are actually made of strings doesn't mean that a theory built around this assumption has no consequences. Suppose you looked at my life from a distance, say from the limited interactions you have with me as a colleague, but not as a friend. You might think, he speaks intelligently, has good credentials, landed this plum job at Caltech—he appears to be a successful, confident guy. But what am I on a deeper level? This is something that, given our relationship, you might not be able to check directly. So you might theorize. At home, do I read Jane Austen novels, tend quietly to my garden, and play the violin? Or do I guzzle martinis and try to keep my neighbor the garbage man from blowing his brains out? There are certainly certain circumstances in which the behavior of the Leonards of the two theories would diverge, and by observing me in such a circumstance, you could infer which is closer to the truth. And so it is with strings. Even if we are not so intimate with nature as to be able to check directly whether particles are made of strings, the question is, can we manufacture a situation in which the observable consequences predicted by string theories and nonstring theories conflict? To be able to propose such an experiment was the string theorists' greatest hope. Unfortunately, no one could figure out how to do it. The theory was just too mathematically complex.

Since string theorists didn't know how to make any testable predictions, they invented another goal for their theory, at least in the short term. It has been dubbed "postdiction." In this approach, rather than making the prediction of some new phenomenon, string theory would provide the explanation of something that was already known, but not understood. For instance, we know the values of many fundamental physical quantities, such as the mass of the quarks, or the charge of the electron, but have no idea why they have those values. String theory had the potential to change that: It promises to produce these numbers from scratch. But no one could figure out how to do that, either.

During the 1970s little of the promise of string theory had been realized. Then, certain inconsistencies were discovered. Everyone, including John Schwarz, figured it would take another mathematical miracle to eliminate these inconsistencies. Schwarz and a tiny group of collaborators believed so strongly that string theory was correct that they began searching for the miracle. To them, the mathematical structure they had already uncovered—for instance, the promise of including gravity—was already a mathematical miracle, and they were ready to allow the theory to lead them forward to the next one. Everyone else simply dropped the theory.

One of the problems with string theory Schwarz did not try to dispel was the problem of dimensions: String theory is not mathematically consistent in just three spatial dimensions. The strings of string theory had length, breadth, and height, but they also required extension into six additional dimensions that don't seem to exist in the real world. Not as bad as my method of infinite dimensions, but these extra dimensions were not an artifact of a mathematical approximation method. According to string theory, the extra dimensions had to be real. String theorists "solved" this problem by adjusting the theory mathematically so that the extra six dimensions were, like the strings, so tiny in extent that they would have naturally gone unnoticed, and, in fact, be virtually impossible to detect.

It was as if we lived in a two-dimensional world, say confined to the surface of the earth, and suddenly a physicist said, hey, look, there exists this extra dimension, up and down, that we have never before noticed. People might ask, how could we not have noticed something so obvious as a new direction? If this "up and down" really exists, I should be able to jump, or toss a ball upward. You can jump, the physicist says, but the dimension is tiny, so your jump can take you only the tiniest fraction of a millimeter upward. So meager is your jump that you'd never even notice getting off the ground.

To a few, string theory's requirement that extra dimensions exist represented a great discovery—like Planck's discovery of the quantum principle, or Einstein's discovery that space and time are intertwined. To those few string theory presented an exciting challenge: Find an indirect but measurable consequence of the extra dimensions (while, in the meanwhile, still working to eliminate the theory's other inconsistencies). But, even at Caltech, most physicists reacted to Schwarz as if he had proposed that everyone move to Nevada to join the secret team studying aliens at Area 51.

Constantine was one of them. I found him sitting at

his desk. He had an inside office—no window. The fluorescent light overhead buzzed. It would have depressed me to have to listen to the buzzing all day. It would have depressed me to have no natural light. Too many things depressed me then, except when I was working. But nothing seemed to ever depress Constantine. He looked tired, though.

"Got to bed at four. Hey—life is tough," he said. He made some gesture with his hands and face that I understood to mean that life is not tough at all. He had been out partying with his American girlfriend, a stunning blond actress named Meg.

I was jealous of him and Meg. Constantine was very handsome, in a Mediterranean sort of way—of slight build, but perfectly sculpted, with alluring eyes and a great smile. He was always tanned, and though he was in his twenties, his hair had just enough gray to lend him an air of sophistication. When he smoked cigarettes it reminded you of one of those ads meant to make it look sexy. At times I had the secret fantasy of running into him in twenty years to find him all white-haired and wrinkled, maybe even a little hunched over. In my fantasy I would be completely unchanged, except for an intangible maturation that greatly enhanced my sex appeal.

I told Constantine I was going to have a talk with John Schwarz.

"Why would you do that?" he asked.

I said, "I thought he might be a good mentor."

Constantine laughed. "Mentor? He can't even mentor himself."

"He seems to take on students."

"Come on. The guy's been here nine years and he still doesn't have tenure. He's not even a professor. He's a research fellow just like you and me." He made another of his Greek—or maybe Italian—gestures, a dismissive motion with his hands of the sort you might make to a busboy to indicate you are done and he can take your plate.

"Well, if he's been here nine years he must have faculty support somewhere. Some kind of pull," I said.

Constantine took his own pull—on his cigarette. He blew the smoke toward the ceiling, then he looked at me with a smile. "He's a mule. He teaches, he takes on a lot of students. Does the work so guys like Feynman can get a free ride."

"Well, with that big load maybe he'll appreciate having another person to work with," I said.

"I'm sure he'll be happy to teach you all about his work. After all, no one else really cares."

"Thanks for your support, Constantine." I walked out of his office.

"What? Did I say something bad?" he asked as I was leaving.

Schwarz's office was around the corner. His door was open. He looked fortyish, and was very clean-cut. He sat at his desk, reading a preprint, which is what physicists call the manuscript for a research paper. Since the journals take so long to actually publish a paper, most current work is circulated and read in preprint form (and these days can be downloaded from the Web). He looked up at me.

"Yes?"

I introduced myself. He smiled, "Yes, I had heard you were a new arrival."

"I was interested in getting to know everyone, and what they worked on."

"I work on string theory," he said, as if it were a household word.

"I thought maybe you could explain a little about your research."

"I don't really have the time," he said.

"Another time, then . . ." I said. "When might be good?"

He got up and walked to the bookshelf. He gathered a half-dozen preprints and reprints of articles.

"Here," he said, "just read these."

He handed me the material and got back to work as if I wasn't there. He had doled out all the words he was willing to spare, and seemed even to be hoarding his supply of eye contact.

Back in my office I licked my wounds. Constantine stopped by and asked a little too cheerily if I was now Schwarz's latest "disciple." I made a gesture with my middle finger that they don't use in either Greece or Italy. But he figured it out.

What neither of us knew was that within a few years the stack of articles now lying on my desk would be revered worldwide as heralding one of the most promising breakthroughs of the century in theoretical physics.

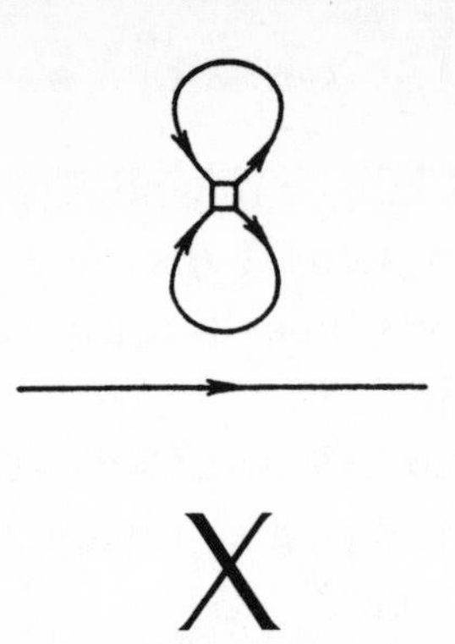

X

IT WAS HARD TO GET a handle on the articles Schwarz handed me, but at least I found myself finally able to focus. I discovered that, despite the dubious reputation of Schwarz and his theory, and his lack of any collaborators on the faculty, he had four or five graduate students working under him, more than any other professor in the department. I spoke with a couple of them when I had questions. They seemed capable. They seemed sane. Didn't they realize that 99.9 percent of physics "experts" thought they were all crackpots?

And why did the rest of the faculty allow so many students to go "astray" like this? Someone, I thought, must be a supporter. Could it be Feynman?

It was a Saturday and the campus was as quiet as the city at daybreak. But this was well past noon, and I was hungry for breakfast. The problem was, though most students lived on campus, on weekends both the Greasy and the Athenaeum were closed. I figured they must eat somewhere, so I went walking outside in search of road kill, or maybe a vending machine. I spotted Feynman a short distance away. I couldn't imagine why he was there, but I took the opportunity to "bump" into him.

"Make any discoveries yet?" he said.

"Right now I'm trying to discover some food. You know where I can eat?"

"I know where," he said. "The difficulty lies in the 'when.' Weekends, the usual places on campus are closed."

We were heading in the direction of the Athenaeum. There seemed to be something going on there. We didn't talk for a while.

"Let me ask you something," I finally said. "Do you think it would be wise to work on a theory that almost everyone else thinks is nonsense?"

"Only under one condition," he said.

"And what is that?"

"That *you* don't think it is nonsense."

"I'm not sure I know enough to tell."

He chuckled. "Maybe if you knew enough to tell, you wouldn't work on it, either."

"You mean maybe I'm too dumb to know better."

"Not necessarily. Maybe you just don't know enough, or haven't known it long enough, to be *spoiled* by what you know. Too much education can cause trouble."

It is true that many of the greatest discoveries in physics were made by "kids" who were roughly the age I was. It was the age at which Newton invented calculus, Einstein discovered relativity, and Feynman developed his diagram technique. Plenty of other advances were made by older physicists, but the most revolutionary seemed to be made by the young. It had been understood among us graduate students that, for mathematical and theoretical physics brainpower, our minds were at their peak. But Feynman seemed to be seeing it differently, as if we go downhill not because of mental decline, but due to some kind of brainwashing. Maybe that's why he avoided learning new things from books or research papers; he was famous for always insisting on deriving new results himself, on understanding them his way. To him, to stay young meant to retain a beginner's outlook. He had clearly succeeded.

"Look," he said. "You've found food."

There was a big buffet in the Athenaeum courtyard. There seemed to be a wedding reception going on. We stopped and gazed at the crowd in their elegant dresses, suits, and ties.

"Yeah, but unfortunately we're not invited."

"I see you are an expert in etiquette."

"What do you mean?"

"I mean, if you are not invited, does it mean you aren't welcome?"

I shrugged. "I usually assume that."

"Then I guess you aren't that hungry."

I thought about it for a moment.

"Well, we're not exactly dressed for it." He had on a dress shirt and slacks. I was clad in shorts and a T-shirt.

"Of course we aren't. What scientist goes to work looking like he's dressed for a wedding? Well, other than Murray." He laughed.

"You'll come with me?" I said.

He grinned. We headed to the buffet. He looked on as I started loading my plate. At first no one seemed to pay much attention to us, but then a man in a tuxedo came up behind us in line.

"Bride's side or groom's?" the man asked.

"Neither," said Feynman. The man looked us up and down. My mind raced, searching for a lie that might minimize my embarrassment. Then Feynman said, "We represent the physics department."

The man smiled, took some salad, and walked away, seemingly unbothered by either the answer or our attire.

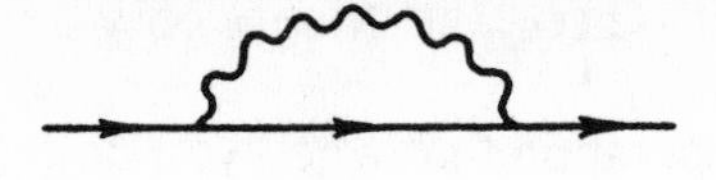

XI

STAYING PLAYFUL, having fun, keeping a youthful outlook. It was clear to me that for Feynman, staying open to all the possibilities of nature, or life, was a key to both his creativity and his happiness.

I asked him, "Is it foolish to become mature?"

He thought for a moment. He shrugged.

I'm not sure. But an important part of the creative process is play. At least for some scientists. It is hard to maintain as you get older. You get less playful. But you shouldn't, of course.

I have a large number of entertaining mathematical type of problems, little worlds of this kind that I play in and that I work in from time to time. For example, I first heard about calculus when I was in high school and I saw the formula for the derivative of a function. And the second derivative, and the third. . . . Then I noticed a pattern that worked for the <u>nth</u> derivative, no matter what the integer <u>n</u> was—one, two, three, and so forth.

But then I asked, what about a "half" derivative? I

> wanted an operation that when you do it to a function gives you a new function, and if you do it <u>twice</u> you get the ordinary first derivative of the function. Do you know that operation? I invented it when I was in high school. But I didn't know how to calculate it in those days. I was only in high school, so I could only define it. I couldn't compute anything. And I didn't know how to do anything to check it or anything. I just defined it. Only later, when I was in the university, did I start over again. And I had a lot of fun with it. And found out that my original definition that I thought up in high school was right. It would work.
>
> Then when I was in Los Alamos working on the atomic bomb, I saw some people doing a complicated equation. And I realized that the form they had corresponded to my half derivative. Well, I had invented a numerical operation for solving it, so I did it, and it worked. We checked it by doing it twice, which is just the ordinary derivative. So I did a nifty numerical method for solving their equation. Everything, well, not everything, but lots of stuff turns out to be useful. You just play it out.

The creative mind has a vast attic. That homework problem you did in college, that intriguing but seemingly pointless paper you spent a week deciphering as a postdoc, that offhand remark of a colleague, all are stored in hope chests somewhere up in a creative person's brain, often to be picked through and applied by the subconscious at the most unexpected moments. It is a part of the creative process that transcends physics. For instance, Tschaikovsky wrote, "The germ of a future composition comes suddenly and unexpectedly. If the soil is ready . . ." And Mary Shelley: "Invention does not consist in creating out of void, but out of chaos." And Stephen Spender: "There is nothing we imagine which we do not already know. And our ability to imagine is our ability to remember what we have already once experienced and to apply it to some different situation."

> Another very interesting and entertaining thing is to ask, if I could change nature in some way, change a physical law, what would happen? First of all, if I would change anything, it would have to be consistent with some other things. And I have to also work out all of the consequences in this modified law to see what would be happening in the world as a result of this thing. It's quite an interesting job. It's a lot of work. And I tried to do that once, I wanted to see how physics would be if it were two-dimensional instead of three-dimensional. Two space dimensions—like Euclid's plane—plus one time dimension. And there are very, very interesting phenomena like the way atoms behave—their spectral lines, for example. I went through a large number of things that are different in two dimensions versus three dimensions. It's very interesting. I have it in a notebook. I had a lot of fun doing that.

By spectral lines, Feynman is referring to the characteristic light that an atom radiates. Adding new spatial dimensions was easy for me to envision. For my dissertation, I, too, had studied how this varies with dimension—all the way from one to infinite dimensions. It was like adding new directions. In one dimension, there is only forward and backward. To get two, you add right and left. For three, up and down. For each additional one, you simply add a new possible independent direction (for some of us, a new possibility for getting lost). It was nice to feel that our imaginations had brought us to envision similar alternative worlds. But I wasn't ready for the strange place he went next . . .

> And then I had fun doing another one. Suppose there were two times. Two spaces and two times. What kind of world would it be with two times?

We are accustomed to events having a temporal order. With two time dimensions—if time must be tracked on a plane, rather than a timeline—there would no longer be a strict order to events. It would be a strange world indeed.

> My son and I discussed that on the beach for a long time. He has a lot of good geometrical imagination. He had made a kind of a model by which we could picture this, so we could figure out just what things would look like. So we could picture and ask ourselves questions. What happens and so forth. That's another game I'd like to play sometime when I have nothing to do.
>
> We do that all the time, ask, "What if?" and start looking at the consequences. But there are so many things you could change, so that unless you have a good reason, you don't bother to change these. It takes imagination to find the right one because if you allowed yourself to make simple modifications like that, there are an infinite number of ways that you could modify things, and it would be very hard to select the right one.

> Someone once said, "What if everything was made out of three particles?"

Feynman is being coy here—the "someone" he is talking about is Murray, and the three particles are his quarks, the particles that are the building blocks for subnuclear particles like the proton.

> Well, then this particle called the K-meson wouldn't fit into the pattern. No good. What if the charges on the particles were nonintegral, though? Ah! That would account for it! Hey, that's nifty. Look, that would produce this! That would explain that! That would explain this thing we never understood before! Big excitement! So now we know that things are made out of three particles that do not have normal charges!

Physicists had noticed long ago that all electric charge seemed to come in multiples of a certain smallest charge. In 1891 Irish physicist George Johnstone Stoney proposed that there existed fundamental, indivisible particles that carried this elementary charge, and coined the word electron. A few years later scientists experimenting with cathode rays observed individual electrons. Since then, no one has ever observed any ion or particle whose charge had a magnitude that was not equal to either 1, 2, 3, or some other integral multiple of the electron's charge. Thus the concept of a "nonintegral," or fractional, charge was very controversial when Murray first proposed quarks. Yet, like the mysterious extra dimensions in string theory, it was necessary for the consistency of his theory.

Cognizant of possible negative reaction, Murray was tentative in his early proposals on quarks. He avoided submitting his initial paper on quarks to the *Physical Review,* fearful of attacks he expected from its editors and referees, and published it instead in a journal of lesser prestige. Feynman, in the beginning, was one of those skeptical of the quark theory. In the end, his own initial hesitation seemed only to increase his admiration of Murray for having developed it.

> To release yourself from the proposition that all charges have to be an integer, and yet everything you see has an integer charge, that took imagination. It takes imagination to say that charges may not be the way you see them all the time. There is a certain conservatism built in. We have established that things are always integral charges, everywhere. Everywhere! So you figure what everything is made out of is also integral charges. It seemed reasonable, and nobody would think of an alternative because it did not seem necessary, and there was no evidence for it.
>
> When you're all finished and you discover something you didn't expect—something that's there that you came across—that looks like it's magic at first! It's fun! It's very interesting. I have investigated many little problems. That's my role.

Listening to Feynman, I was inspired. Why not release myself from the idea that space-time had four dimensions? So what if string theory required six more? It was a "what-if" I figured deserved more investigation.

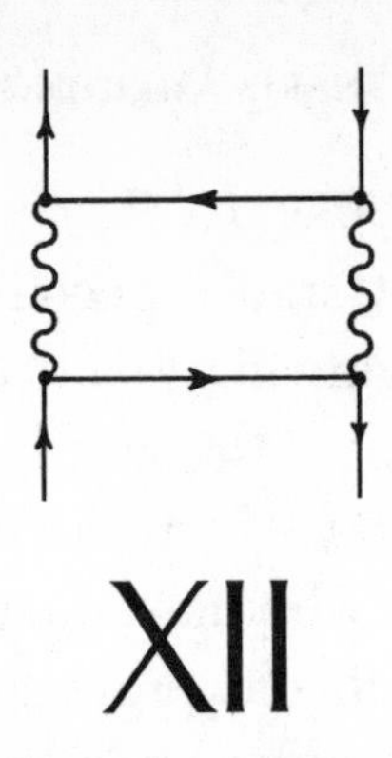

XII

SPRING WAS NEAR. It's a nice season in Pasadena—warm weather, but not yet hot, and less rain than winter. A time to enjoy the blue sky, palm trees, and a clear view of the San Gabriel Mountains still blanketed in green. Somehow, somewhere, Ray finally met a girl he liked, or, more to the point, who liked him. The only problem, according to Ray, was that she lived in the state of Washington. Bellevue, to be exact. I saw additional problems. Like the fact that he had decided not to tell her he was a garbage man, only that he worked for the city. And that the only thing they seemed to have in common was that they were both great at math, at least elementary math. But since Ray happened to hate math, I didn't necessarily see the math connection as a plus. Still, he seemed pretty serious about her, and I was happy for him. He was even thinking of moving to be near her. She did some work for a small software company up there called Microsoft. He thought maybe she could help him get a job. I, of course, selfishly hoped he would stay put.

Since I often spoke to Ray about the Caltech physics department, and especially of, as he always put it, "that guy Feynman," Ray decided he wanted to see the place and meet the guy. I agreed, though not without trepidation. Introducing a loquacious cannabis aficionado who hates math but loves talking philosophy to a gruff old professor who likes math, hates talking philosophy, and is fiercely protective of his time is not without risk. But Ray and I were friends, so I agreed to do it.

Ray often asked me what physicists did, and why they did it. One time I answered him by reciting an Einstein quotation I had read in *Zen and the Art of Motorcycle Maintenance:* "Man tries to make for himself, in the fashion that suits him best, a simplified and intelligible picture of the world . . . and thus to overcome it. . . . He makes this cosmos and its construction the pivot of his emotional life in order to find in this way the peace and serenity which he cannot find in the whirlpool of personal experience."

"That's just like Einstein," Ray had said. "His head was way up in the clouds. What I wanna know has to do with planet earth. I wanna know . . . what—do—you—do, and why—do—you—do—it?" He said it as if repeating the question slowly and with emphasis on each word somehow gave it another meaning. If it did, it went over my head. But I thought a visit to campus might provide that picture that was worth a thousand of my ineffectual words.

On the way over, I tried my detective metaphor.

"It's a lot like Sherlock Holmes, or Rockford, depending on your personal style, of course. The first thing is, you have to choose a problem."

"Like choosing a crime to work on."

"Right. Except detectives are assigned cases. Physicists have to choose for themselves."

"Is there the equivalent of the FBI Ten Most Wanted list?"

"Sure, there's problems everyone thinks are important. But you've got to be careful—a lot of people are working on those. It's better to find a problem that only you realize is important to solve—that is, if you are right about its being important."

"And then you look for clues."

"Yeah, but it's all in your head. You mull over the possibilities, come up with ideas—leads. Then you chase down the leads by fiddling with the mathematics. To figure out if your idea has the consequences you thought, or not. Often that's not so easy, because you don't know how to do the math. Am I making sense?"

"Only in some abstract and totally superficial way."

I smiled. "Sounds like progress."

After a quick stop in my office we stepped into the hall and walked around the corner. There were already a few graduate students milling around outside the seminar room. Physicists thrive on discussion. They'll talk physics anywhere, just as anyone else talks about sports or the weather. It gives them a chance to cross-pollinate. That's how Schwarz made his greatest breakthrough—well, what *he* considered a breakthrough, anyway. He had been casually chatting with Michael Green in the cafeteria at the European Center for Nuclear Research in Switzerland a couple of years before, when suddenly, together, they realized that string theory was also a theory of gravity. Had they found that, say, quantum chromodynamics could be extended to include gravity, it would have been front-page news around the world and a sure Nobel Prize. But, virtually no one thought string theory was correct. The fact that this incorrect theory might also include a description of gravity did not arouse much excitement among the few who even bothered to listen.

I had to admire Schwarz—massive rejection didn't stop him from pushing his theory at every opportunity.

Today he was giving a seminar on his work with Green. Whenever a faculty member or student has found something worth explaining, and often when he hasn't, the seminar room is the place to let your colleagues know en masse about your work. In Schwarz's case, en masse would probably mean only a handful who bothered to show up, but Schwarz always took it with a smile. And he seemed to give more seminars than anyone else in the department.

I admired him for something else, too. Schwarz, like me, had gone to Berkeley. His Ph.D. advisor there, in the sixties, was a fellow named Geoffrey Chew, who was the leading figure behind another very ambitious approach called S-matrix theory. The aims and philosophy behind S-matrix theory were similar to those behind string theory, and for a few years it was the hottest thing around, but it didn't pan out. Chew, however, didn't drop it, and for decades he worked, like Schwarz, snickered at and virtually alone. Chew got nowhere and ended his once brilliant career in oblivion. For Schwarz to work in Chew's shadow, to seem to be repeating his history, and still to move forward smiling, I thought, showed great character.

I knew Ray wouldn't understand a word of the seminar, which would put him only slightly behind me, but I figured since he kept asking what it was we *really* do all day, I might as well give him a taste of it.

Only about ten people, half of whom were Schwarz's graduate students, showed up for the talk. But shortly before the talk began, joining the group loitering outside the seminar room were both Murray and Feynman. It was the first time I had seen them both attend a seminar, and I figured it could mean fireworks.

Some years earlier, when it was more common to see Feynman and Murray both attend, seminars at Caltech had the reputation of being brutal events. Murray might challenge you incessantly, even on the tiniest point. Or worse, if he thought what you were saying was of little importance or interest, he might pull out a newspaper, and read it in conspicuous boredom. Feynman, too, was always brash and unwilling to accept wrong or sloppy thinking, and he seemed to revel in playing cat-and-mouse. To Feynman physics was a show, and if you didn't satisfy him with your answers his response was sometimes to get up, announce his opinion, and march out of the room. The combination of Murray and Feynman was so intimidating that at least one future Nobel Prize winner hesitated to lecture at Caltech.

As we walked over Murray was speaking to a visitor who apparently had just come from Montreal. Only Murray insisted on pronouncing the city's name as the natives did, "Mon-ray-al."

Feynman turned to face Murray. "Where?" he said.

"Mon-ray-al," repeated Murray.

"Where's that?" said Feynman. "I've never heard of Mon-ray-al." He exaggerated Murray's pronunciation for effect.

"I've observed that there are many well-known cities whose names you don't seem to recognize," said Murray.

"Logically speaking, that means that either I'm an ignoramus . . . or you say them in a funny way."

"Not true," said Murray. "Logically speaking, it could also be both." Murray was always a stickler for precision.

Feynman smiled. "Well, we'll just have to let everyone draw their own conclusions."

Murray smirked and walked into the seminar room. Feynman found teasing Murray to be fun and games; Murray always let it upset him. I quietly pointed Feynman out to Ray.

"Who was the other one?" he said.

"Murray Gell-Mann."

"Oh, the quarks guy."

"Yeah, the quarks guy."

"Do they always talk to each other like that?" he asked.

I shrugged. I rarely saw them together.

"They remind me of my mother and father," Ray said.

As the seminar began Feynman yelled out, "Hey Schwarz, how many dimensions are you in today?"

It wasn't the only time I heard him utter that gibe, referring to the extra dimensions required by string theory. But it was always good-natured. This meant something, because Feynman's quips did not always possess that quality. So I didn't feel it necessarily showed where he stood on the subject. I felt a little tense, standing there, waiting with Ray. I was ready to watch a fight—would Feynman and Murray team up against Schwarz, or would they end up somehow battling each other? I was a little embarrassed to have brought Ray, the way you might be embarrassed to have a friend hear your parents argue.

Schwarz smiled and began his talk. He seemed at ease. He even wove in a few jokes. They received hardly a chuckle. Years later Schwarz would tell me with amusement how, after he became famous, similar quips would bring roars of laughter.

Feynman and Murray listened respectfully, and asked only a few technical questions. There were no derisive comments.

A few minutes into the talk I looked over at Ray. He was asleep.

At the tea and cookies in the back of the room after the lecture I introduced Ray to Feynman. I had warned Ray not to be too aggressive. And for God's sake not to ask questions of a psychological or metaphysical nature. Feynman has doctor's orders not to discuss metaphysics, I had told him. He had given me a weird look, but I was confident he'd be on his best behavior. Feynman turned to me.

"So, did the seminar teach you anything useful about that 'nonsense' theory you were interested in?" he said.

"You mean you knew all along it was string theory?"

"It's the only nonsense theory we've got going in this department," he said.

"If the theory is nonsense," Ray asked, "why are you here?"

Feynman grinned. "I came for the cookies."

We drifted into the corridor outside the seminar room. At that point the visitor from Montreal, who had been eavesdropping, stepped over.

"I don't think we should discourage young people from investigating new theories just because they are not accepted by the physics establishment," he said.

Something about his challenging tone made me feel this guy would be at home addressing a Berkeley rally against cultural imperialism. But Feynman took it well.

"I'm not telling him not to work on something new," Feynman said. Then he looked at me and said, "I'm just saying, whatever you choose to work on, be your own worst critic. And then don't do it for the wrong reasons. Don't do it unless you really believe. Because if it doesn't work out, you could end up wasting a lot of time."

The visitor said, "Well, I have been working on my own theory for twelve years."

Feynman asked him what theory that was. The man described it briefly. He seemed peeved at the end that no one was impressed. I felt that just for listening politely we should have all been awarded a prize from the give-dumb-theories-equal-time movement, of which I was certain he must have been a member. He seemed to sense this, for he added, "It took the physics community years to accept Einstein. It is taking them years to accept Schwarz. I don't mind if it takes them years to accept my work. It's really a compliment. And it'll make it all the sweeter when recognition does come."

I didn't think the fellow's attitude would go over well with Feynman, but he seemed to be listening intently. And when the fellow finished, Feynman nodded politely, as if he had just learned something.

Then he looked over at me and said, "That's exactly what I mean about wasting your time."

The visitor walked away in a huff. Ray said to Feynman, "How could you say that to him, man? That's cold."

I elbowed Ray.

Feynman said, "You don't like what I told him? Why not? He wanted recognition. I gave it to him. I recognized him as a pompous ass."

Just then Helen appeared down the hall. She was holding some mail, apparently Feynman's. She made a gesture that I took to mean she would leave it in his office. He nodded. Then, spotting me, she called me over. I gave Ray a warning look that said, "Watch what you say!" He gave me a look back that said, "Moi?" I was worried about exposing Feynman to Ray untended, but, when Helen called, you obeyed.

When I finally got back from her office around the bend, the corridor was deserted and Ray and a few butter cookies was all that was left in the seminar room.

"How did it go?" I asked. "Will he ever speak to me again?"

"Relax," he said. And then, "You need some pot."

"Ray, shut up!" I looked around to be sure no one was within earshot. What I didn't know at the time was that Feynman had himself tried marijuana—and even LSD.

"Don't worry, it went fine. We're buddies. Hey, you never told me he had a Nobel Prize."

"He told you that?"

"Yeah."

"I heard he never talks about that. He thinks the Nobel Prize is by its nature unfair. And a big distraction. A false God, so to speak. He told me when the first reporter called him in the middle of the night to tell him he had won it, he told him to call back at a decent time and hung up."

"Well, maybe he feels that way. But maybe he's also proud. That would be human, wouldn't it? Maybe he just doesn't open up to you like he does to me."

"So now you and he are best buddies, I suppose."

"Well, you know what else he told me? He finally explained to me what you physicists all do, and why you do it."

"He did?"

"He did."

"What did he say?"

"No, no, no," he said. "You don't get off that easy. Ask him yourself. Or even better, find your own answer."

"Now *you* sound like Feynman," I said.

"Well, we did see eye to eye on certain things."

I let it go at that. But I figured, one way or another, I'd get it out of Feynman.

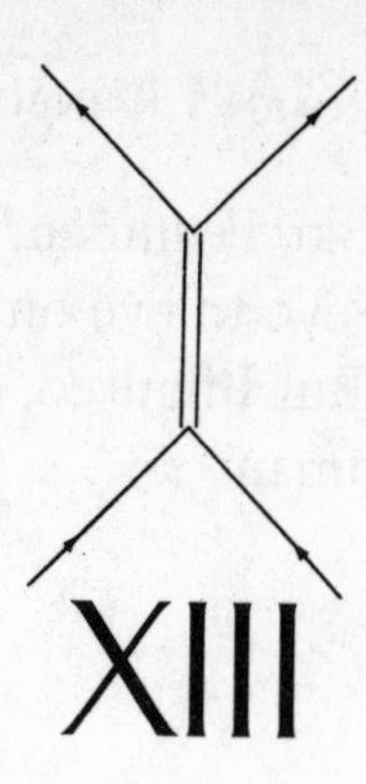

XIII

IN 1988 A FORMER CLASSMATE of mine from Berkeley started writing a text on string theory that is now a standard reference for physics graduate students. He projected completing the book a year later in June 1989, "plus or minus one month." It is not uncommon that books are finished late, but this book wasn't published until 1998. It had taken eleven years, over ten times the projected period. Why? String theory is hard. There are famous stories about how few people understood relativity and quantum theory in the early days, and even in more recent ones. But it is safe to say that, even today, nobody understands string theory.

Most new theories are demanded by nature. They grow out of new physical principles or experimental facts that need to be explained or accommodated. String theory did not arise that way. String theory was like penicillin, uncovered by accident. Theoretical physicists are still searching for the new physical principle string theory presumably represents. Experimental physicists are still searching for an experimental consequence they can test in the lab. Physicists who study it are like paleontologists, patiently digging and scraping at it, as if they are uncovering a giant skeleton of unknown origin.

It all began in the summer of 1967. Murray, who hadn't yet received his Nobel Prize, was giving a lecture at the Centro Ettore Majorana in Erice, Sicily. He was speaking about some issues in S-matrix theory, that theory championed by Schwarz's Ph.D. advisor, Geoffrey Chew. That theory that never panned out. In the audience was an Italian graduate student (then working in Israel) named Gabriele Veneziano. Murray, ever the classifier, ever the Greek, was discussing some striking regularities in data pertaining to the collisions of protons and neutrons. Veneziano was intrigued. It took him a year, but he eventually found a simple mathematical function that magically described the regularities. The word magical here is not used cavalierly: Veneziano did not employ any theory of physics to derive the function; he simply discovered the mathematics that worked. It took a couple more years for physicists to propose a reason *why* it worked. The why was first presented in 1970, in the work of Nambu and Susskind, who figured out that Veneziano's mathematical function would arise from the underlying theory if you modeled the protons and neutrons, not as point particles, but as tiny, vibrating strings.

That seemingly simple idea, it turned out, was far richer, and far harder to implement mathematically, than anyone might have guessed at the time. And though it was a physical model of what particles are made of, it wasn't a physical *principle*, like the constancy of the speed of light, that could guide you in your thinking as you sorted through all the possible ways to develop the theory. That's another reason string theory is so hard.

After my two attempts at gently raising the issue of string theory, one afternoon I stepped into Feynman's office to ask him what he really thought.

"Can we talk a little about string theory?" I asked.

"I don't want to talk about string theory. I don't know much about it." He looked back down at his work. "You wanna talk about string theory, go talk to Schwarz."

"I did."

"Then go talk some more. I'm working."

"It's hard to understand, and I'm trying to decide if it is worth the effort."

"Like I told you, only you can decide that."

"Don't you think there are aspects of it that seem very promising?"

"Promising? What does it promise? Does it promise to tell you the mass of the proton? No. What does it promise to tell you?"

"Well, no one knows how to extract any quantitative predictions yet, but—"

"You're wrong. It does make a quantitative prediction. Do you know what that is?"

I looked at him. My mind was a blank.

"It requires that we live in ten dimensions. Is it reasonable to have a theory that requires ten dimensions? No. Do we see those dimensions? No. So it rolls them up into tiny balls or cylinders too small to detect. So the only prediction it makes is one that has to be explained away because it doesn't fit with observation."

"I know . . . there's a lot to be worked out. But what intrigues me is that string theory has the potential to unify all known forces of physics into one theory. Even gravity."

He looked at me with a strange expression. The kind you might expect if you were making small talk with a Catholic bishop and casually inquired about his wife and kids.

"A unified field theory. Isn't that what we all want?" I said.

"I don't *want* anything. Nature has nothing to do with what *I want!* How do you know there's one unified theory? Maybe there's four theories! Maybe there's a theory for each force! I don't know. I don't tell nature what to do. Nature tells me. This whole discussion is pointless! It's getting on my nerves! I told you—I don't want to talk about string theory!"

This last part got loud. Plus he was waving his arms. I was taken aback. First, because I thought the reason we all did physics was our passion for the beauty and elegance of nature, and four theories didn't seem very elegant to me. And second because from the expression on his face, I was afraid he might get up and bite me. I figured it was time to make my exit.

"Look, I'm sorry. I just wanted to get your take on it."

"My take? My take is that you hit a dry spell, and now you're scrambling, trying to find something to work on."

"Is that wrong?" I asked.

"What's wrong is coming to me to talk string theory."

"Your opinion matters to me."

"Like I told you before, what should matter to you isn't my opinion. It's *your* opinion."

"I'm sorry I bothered you," I said, and started to leave.

"Look," he said, "selecting a research problem isn't like climbing a mountain. You don't do it just because it is there. If you really believed in string theory, you wouldn't come here asking me. You'd come here *telling* me."

I felt like a little kid who had just been scolded by his dad. Back in the corridor, I got scolded again, by Mom. I ran into Helen. Though she was the secretary for the entire floor, she worked mainly for Feynman and Murray. A thin, middle-aged woman, she had the grit to stand up to both of them, and tons more grit than she needed to handle me. She was sporting a major frown.

"What did you say to piss off Professor Feynman?" she asked.

I shrugged.

"You know you shouldn't interrupt him when he is working."

"I guess I just tried to engage him on the wrong topic."

"Philosophy?" she asked.

"No, string theory," I said.

"Oh, God, that's just as bad."

"Can I ask you a question?" I said.

"Maybe," she said. "What is it?"

"If everyone is so skeptical about Schwarz's work, why is he still here after nine years?"

She gave me a look. I didn't know if she meant, "You mean you don't know?" or "Why do you care?" But after a moment, she said, in a lower voice, "He has someone who looks out for him."

"Oh? Who's that?" I said.

She said, "Murray."

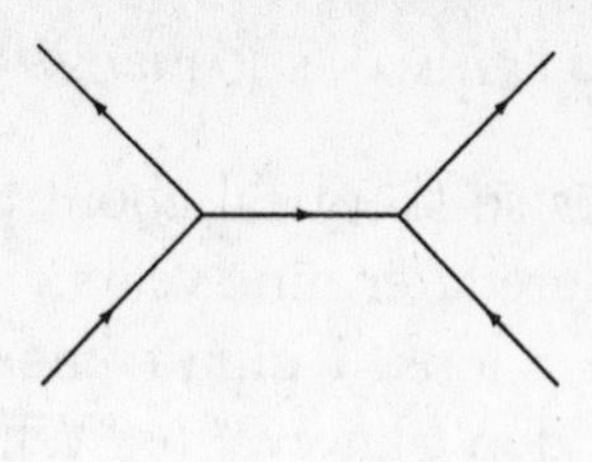

XIV

A COUPLE OF DAYS afterward I was at the office late in the day. Earlier, Constantine had told me that Murray had been estranged for years from his daughter, who had joined what became the Marxist-Leninist party of the U.S.A., and was a big fan of Albania. Apparently, although Murray derided Reagan by calling him Ray-gun, Lisa had crossed a line by singing songs like "Down with Ronald Reagan, Chieftain of Capitalist Reaction!"

I sat at my desk thinking there were ironic parallels with Lisa's politics and Murray's underground support for Schwarz's subversive theory. For in their own way, Lisa's politics were not any farther from the mainstream than string theory, or for that matter, than Murray's earlier discovery/invention of fractionally charged quarks.

Did daughter inherit from father the ability, for theory's sake, to be unmoved by seemingly obvious data—like the absence in our world of string theory's extra dimensions, or the absence in Albania of certain amenities, like food, clothing, and shelter? Did they share a genius (or curse) of being able to see through the façade of reality to a more fundamental truth?

My ruminations were interrupted by Murray, whom I could once again hear yelling through the wall. It was disquieting, but that didn't bother me, for my office, anyway, was far too quiet for my taste. What did bother me, with communism still fresh on my mind, was the thought of the poor downtrodden individual on the receiving end of his tirade. I decided if Helen could bring it up to him, so could I. I would tell him a thing or two.

As I stepped into the hall, my heart raced. After all, Murray needed Helen. Along with him and Feynman, Helen seemed to me to form the soul of the department. I, on the other hand, was expendable. Murray could squash my career with nary a thought. I imagined the worst, that I would find myself barred from the department supply of paper and chalk. Or that my office would be moved to the boiler room—or, perhaps with Lisa's help, Albania. By the time I got to Murray's door, though, the yelling was over. I was relieved.

I noticed the door was open a crack. This was unusual. Both Murray and Feynman usually kept their doors closed. It helped cut down on interruptions from the students and junior faculty like me. It also helped keep out the occasional nutcases who always plague the top schools. They'd come by with their new discoveries. Particles faster than light, or the universe is a pancake and we are the syrup—it didn't matter what they believed, they always saw themselves as the new Einstein. If you're unlucky enough to encounter one of these undiscovered geniuses, it could be a couple of hours down the drain. You had to be careful how you rejected them, because it sometimes turned out they were armed. At Berkeley, there had been a fellow who responded to rejection by hanging out outside the physics building with a knife. My Ph.D. advisor told of a guy at Columbia once who returned with a gun. His professor was out, so he killed his secretary.

I looked through the crack in Murray's door. I expected to see him leaning back in his chair, smiling over whatever victory he had no doubt just achieved. But what I saw instead was a man who looked broken, his elbows on his desk, his head resting in his hands. His face was full of agony. I had lost all desire to yell at him. Instead, I felt bad for him. I didn't know why he was upset. The next day I returned to my Greek Oracle, Constantine, for the answer. He told me Murray's wife had recently died of cancer.

I decided to stop eavesdropping and step away. But it was too late. He had spotted me.

"Can I help you?" he said.

I stood there, busted. What was I going to say? I came over to tell you to stop yelling at people, but then I decided to spy on you instead?

"Oh, hello. Come in," he said, recognizing me through the crack.

I opened the door and stepped in, feeling awkward.

He added, "I want to thank you again for giving me your brother's wonderful book."

A couple of years earlier, while still in high school, my younger brother, Steve, had written a book on the birds of the Chicago area. Murray was an avid bird watcher and conservationist. He could rattle off the identifying characteristics of various birds with the same comfort he had speaking Upper Mayan. He could probably rattle off the characteristics of birds *in* Upper Mayan. So, when I moved in next door, I had given Murray a copy as a kind of reverse housewarming gift.

"That was very nice of you," he continued.

"My brother was excited when I told him you were reading it."

Murray smiled. "So, what can I do for you? I saw you at John's strings seminar the other day."

It seemed like an opportunity.

"I was wondering . . . what are your views on string theory?"

"I think it is very promising."

"Promising in what way?" Given my experience with Feynman, I was proceeding with caution. I didn't want to say anything stupid. But I just had. How could anyone who had read anything about string theory not know why some people thought it promising? Feynman might have skewered me for that, but Murray didn't seem to mind the question.

"It could be the theory that unifies all the forces of nature. To have a single theory of the gravitational force, the electric force, of all forces, that was Einstein's dream. Doesn't that inspire all of us? Imagine, a single, simple formula that explains the great multiplicity of particles and all their interactions!"

"Yet people are very skeptical."

"They have a right. But it is still worth pursuing. Look, when I first brought John here almost ten years ago, we didn't even know the connection between gravity and strings. Back then, I didn't know *what* strings would be good for. But I knew it would be great. It was too beautiful not to be. Obviously, not everyone necessarily saw it that way. Then, when John Schwarz and Michael Green found the connection to gravity, it was heartwarming. It made me proud and happy to have John here at Caltech. Still, some influential people don't understand. There is some crazy opposition. Even hostility."

"I guess people don't see its connection to reality," I said.

"It's because research on string theory is proceeding in the most unorthodox of steps. Creating this theory is a process of discovery, not invention. They are looking for something that is *there,* not *creating* something to fit experimental data. Progress is slow. But the hope is that people are piecing together a unique, self-consistent theory. That's why I support them. I have a gut feeling there is something there. Let's just say I'm maintaining a nature reserve for endangered theories."

As I would learn later, Feynman had no objection to the idea that a theory like string theory was already "there," waiting to be unearthed, as Murray believed about string theory. But Feynman felt that only a principle or observation of nature could lead us to the right theory, not a scientist's desire for unification. It was his Babylonian approach—worship the phenomena, not the explanation.

So Feynman scorned string theory, Murray championed it. That was Feynman and Murray—attracted by each other's genius, repelled by each other's philosophy, held in orbit by the balance. Somehow, I couldn't imagine either of them staying put without the other. When Feynman died, it seemed to me, Murray would hurtle out of orbit, the way the moon would if the earth suddenly disappeared.

The goal of science may be to describe reality, but as long as science is carried out by human beings, human qualities will affect the description. The Feynmans will stay close to the data, the Murrays will be guided by their philosophy, their need to classify nature neatly and cleanly. In the end, one or both may succeed, and if they both do, then a peacemaker will show how their theories match up, just as Freeman Dyson had done for Feynman diagrams. Just as in quantum mechanics energy can be viewed as either particles or waves, different visions may both be correct, nothing but different views of the same many-faceted miracle, nature.

Murray proved to be a good conservationist. Though there had been considerable pressure not to renew Schwarz's position, he had recently instead been given a minor promotion, to senior research fellow, and a new contract for three more years. It still wasn't what Murray had wanted for him—a regular tenured position—but it worked for the time being.

When I found out about Murray's wife's death, I admired him for having the focus to do even that much for John. Margaret had been sick for over a year. It was a hopeless form of cancer, colon cancer that had spread to her liver.

At first, Murray had approached the cancer in much the way Feynman approached his—he learned everything there was to know about it and became integrally involved in deciding the treatment. In the end, their approaches differed. Feynman as usual stuck close to the data—that there was little more they could do for him. But Murray had a hard time accepting that with his genius and with all the resources of modern science available to him, he couldn't save Margaret, his only true friend. Even after being told there was no hope, he desperately tried to keep her alive with experimental treatments in the hope that in the interim, a cure would be discovered.

And in the midst of it all, he managed to keep John Schwarz afloat at Caltech.

Constantine told me that the prevailing view was that in the short time since Margaret's death, Murray had mellowed. He didn't yell as loudly as he used to, or as often. He didn't quite seem to be the same Murray, Constantine said. I had never known the "old Murray," but as I observed him over the next year, I did detect a gradual softening. I never again heard Murray yelling through the office wall. I wondered, was it merely that his energy was sapped, or did it go deeper? Had he, through his loss, somehow found a better way to live? In time, I grew to feel sorry for him. Not because he no longer felt the need to rant and rave, or to constantly prove himself superior, but because, for the first fifty-two years of his life, he had.

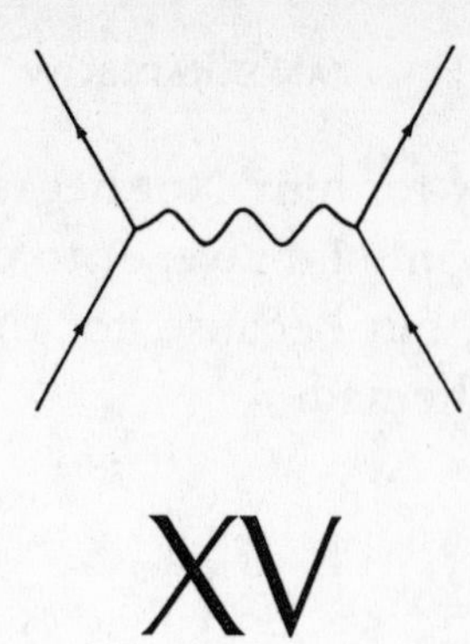

XV

Constantine and I walked down the olive walk in the late afternoon. The campus was quiet. It had rained through the night and morning, but the rain had recently let up. The branches of the olive trees glistened in the emerging sunlight. A while back Feynman had suggested I drop by to see an undergraduate who lived in a nearby dorm. I finally decided to go, and I had grabbed Constantine to join me.

His eyes were red. Another long night with Meg. Drinks at some "in" bar in Hollywood. Then his Fiat broke down in the rain. A great car, unless you needed to be somewhere. It worked for Constantine, though. A ride home in a tow truck, and then he and Meg made love all night. Constantine had said a few times that Meg and he didn't seem to mesh on an intellectual level, but apparently those other levels made it okay. To me, they seemed made for each other, like cover models from *Cosmopolitan* and *Cigar Aficionado.*

I was feeling lonely, and glad he'd agreed to come

along. Always ready for an adventure, that was Constantine.

"What's so special about this guy that Feynman sends you there?" he asked.

I shrugged. All I knew was Feynman said it would be interesting, or, as he would pronounce it, IN-ter-ES-ting. Apparently the undergrad had a collection of spiders. I figured it had to be a pretty good collection to make it worth a special look.

Constantine walked gracefully along the wet sidewalk. He didn't get a drop of water on his elegant Italian shoes. I accidentally stepped in a deep puddle and doused my sneakers. There must have been a pothole in the concrete. As I shook the water off my foot, Constantine asked if I wanted to collaborate with him on his research.

"Forget string theory," he said. "And forget trying to solve quantum chromodynamics with mathematics. Computers, that's the answer. Computers are the future. You want to be a success, get in on it now."

Constantine worked on quantum chromodynamics, but he belonged to a growing number of computer physicists who worked in an area called lattice theories. Since the equations of quantum chromodynamics apparently couldn't be solved by humans, their approach was to have a computer solve them. And since no computer, no matter how fast, can handle the infinity of points in the space-time continuum, lattice theorists had to rewrite the equations in terms of a finite lattice of points—hence their name, lattice theorists.

Constantine's proposal caught me by surprise. He

sounded a bit like Ray talking about his girlfriend and her work up in Bellevue. "You'll see," Ray had said, "someday computers will be everywhere. They'll be like HAL in *2001*."

"Maybe," I said, "but will they be able to pick up the garbage?"

"No, I figure my job's safe," he said. "But I bet they'll be able to smoke pot."

"That'll be a sad day," I said.

"Not really," said Ray. "They won't replace the human. They will augment him. With HAL getting stoned at your side, the party will just be that much better."

I had had a little experience programming computers, but I didn't see them improving any parties. Nor did I see them as the panacea for unsolvable theories. I liked Constantine, but I didn't really believe in his approach. Getting answers from a computer was like getting them from a black box. I felt they gave solutions—numerical results—without providing the understanding you get when you solve or approximate the equations yourself, mathematically. Because of this, I didn't even trust computer solutions. I had never mentioned any of this to Constantine, and I didn't see what good it would do to say it now. Plus, I figured that the fact that I didn't believe in the approach didn't mean it wasn't the right one, or even that I shouldn't do it. I had to weigh against my personal intuition the fact that lattice theories were a lot more "in" than string theory, and a lot more conducive to a future tenured position. And I'd probably like working with Constantine.

"Hey," he said, reading my hesitation, "we calculated the mass of the proton. That's something no one can do using straight mathematics."

He was right. The mass of the proton was a simple thing for experimentalists to measure, yet theoretically the proton mass depended on the quarks inside it and their interactions via the strong force, and it was one of those problems in quantum chromodynamics that no one knew how to solve. Constantine had made quite a splash doing it via the computer: Even many computer skeptics were amazed at the accuracy of his answer.

He winked at me. "Got me to Caltech, didn't it?"

We found the room, and Spider Guy answered the door. He was thin, and wore a Caltech T-shirt that was several sizes too big. He had a large room that was bright with the fresh sunlight, but I doubted he appreciated that. He'd have been just as well off in a cave, I thought. The same went for what, from the look of things, were the room's principal occupants—several hundred spiders.

The room was jammed with card tables, arranged to cover the floor space with mathematical efficiency, but not for human convenience. There was hardly room to walk among them. On the card tables were rows and rows of small plastic cups. Each one contained a spider, or at least a spiderlike bug. Big spiders. Tiny spiders. Hairy spiders. Bald spiders. Here and there were spiders he announced were venomous.

"They can't crawl out," Spider Guy said. "See." And then he tilted one of the cups to demonstrate how it was too slippery for the spider to climb up the side.

Was it the wax coating? Had he sprayed it with Pam? I didn't know, but whatever his trick, it worked. Thank God for that, I thought. Then I wondered what would happen in an earthquake. There had been a 7.2 up near Eureka a year ago last November. Constantine's thoughts were apparently less theoretical.

"Hey," he said after checking out the collection, "where do you sleep?"

And it struck me—there wasn't a bed, or even a chair, in the room. Just these spider tables.

"Under the tables," said Spider Guy.

"The girls must love that," said Constantine.

"Oh, I go to their place for that," said Spider Guy.

Given his interests and the paucity of female students at Caltech, I marveled that he got any of "that." Or that he especially wanted it. He seemed to be in love with his spiders.

We left.

"I wonder why Feynman sent you to see *that*?" Constantine said.

"I don't know. But he was right. It sure was interesting," I said.

"In a sick way," he said.

I shrugged. "I thought he seemed pretty happy," I said.

"Hey—sometimes sick people are the happiest. They're too sick to know how unhappy they should be."

He stopped to light a cigarette.

"Schwarz is probably happy, too. He probably sleeps under a pile of strings," he said. He slowly exhaled a billowing cloud of smoke. Suddenly I wanted a cigarette. It seemed to bring him such deep satisfaction. "Let me know if you want to learn lattices," he said. "I'll promise you one thing . . . you won't have to sleep under a table of spiders—or strings."

With that we kept on toward the physics building. Then I spotted Feynman in the distance. I had spent the last couple of days on the lookout for Feynman, hoping to manufacture a natural way to bump into him and see if he would still talk to me. I told Constantine I'd see him later. I walked over toward Feynman.

When I got to him, Feynman was gazing at a rainbow. He had an intense look on his face, as if he were concentrating. As if he had never seen one before. Or maybe as if it might be his last.

I approached him cautiously.

"Professor Feynman. Hi," I said.

"Look, a rainbow," he said without looking at me. I was relieved that I didn't detect any residual annoyance in his voice.

I joined him in staring at the rainbow. It appeared pretty impressive, if you stopped to look at it. It wasn't something I normally did—in those days.

"I wonder what the ancients thought of rainbows," I mused. There were many myths based on the stars, but I thought rainbows must have seemed equally mysterious.

"That's a question for Murray," he said. I eventually tested Feynman's theory on this and asked Murray. Sure enough, I discovered that Murray was an encyclopedia when it came to native and ancient cultures. He even collected artifacts. I learned from him that the Navajo people saw the rainbow as a sign of good fortune, whereas some other Indians saw the rainbow as a bridge between the living and the dead. I didn't quite get the names of those Indians because Murray pronounced them in a manner that was so authentic it was unintelligible.

"All I know," Feynman continued, "is that according to one legend angels put gold at its ends and only a nude man can reach it. As if a nude man doesn't have better things to do," he said with a sly smile.

"Do you know who first explained the true origin of the rainbow?" I asked.

"It was Descartes," he said. After a moment he looked me in the eye.

"And what do you think was the salient feature of the rainbow that inspired Descartes' mathematical analysis?" he asked.

"Well, the rainbow is actually a section of a cone that appears as an arc of the colors of the spectrum when drops of water are illuminated by sunlight behind the observer."

"And?"

"I suppose his inspiration was the realization that the problem could be analyzed by considering a single drop, and the geometry of the situation."

"You're overlooking a key feature of the phenomenon," he said.

"Okay, I give up. What would you say inspired his theory?"

"I would say his inspiration was that he thought rainbows were beautiful."

I looked at him sheepishly. He looked at me.

"How's your work coming?" he asked.

I shrugged. "It's not really coming." I wished I was like Constantine. It all came so easily to him.

"Let me ask you something. Think back to when you were a kid. For you, that isn't going too far back. When you were a kid, did you love science? Was it your passion?"

I nodded. "As long as I can remember."

"Me, too," he said. "Remember, it's supposed to be fun." And he walked on.

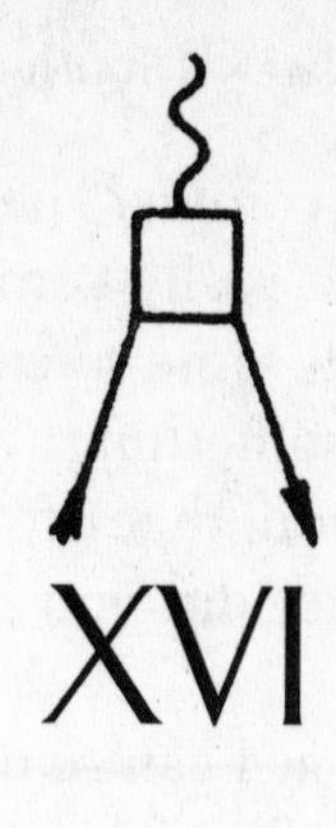

XVI

IN THE BRIEF WINDOW of time I knew Feynman, he had an exaggerated effect on my life. I wasn't sure why. I knew he wasn't going to be any kind of mentor. Feynman avoided all departmental and administrative affairs, and did little to help his own postdocs or students. He would even have Helen send an unusual form letter to all junior physicists he worked with two years after they had left Caltech. The letter said he would no longer write them letters of recommendation because for the past two years he hadn't been following their research. He was diligent in avoiding any activity that he did not find IN-ter-ES-ting. He could be abrupt and abrasive, yet I never lost any of the instant affection that came automatically the first time I met him. Why?

Back then, I did not know the answer. Today, as the father of two young children, I recognize the attraction. Even after the ups and downs of the fifty or so years of adulthood, even in the process of dying, Feynman was still a child. Fresh, gleeful, playful, mischievous, curious . . . IN-ter-ES-ted. Add a few hairs, subtract a few wrinkles, give him his health, and you'd have the same Feynman who yelled fake curses in made-up Italian to scold offending drivers in Brooklyn fifty years earlier.

Hanging around a grown kid like Feynman made you question things. Like all the things we do in life because we have to do them—or at least we *think* we do. Sitting through boring meetings with colleagues or customers or clients when we'd rather be outside staring at a rainbow, or managing our careers along some path for which we have no passion merely because it is supposed to be the road to success. Like my young boys today, Feynman was startlingly honest with people, including himself, and you couldn't make him do anything he didn't want to do, at least not without grumbling. In contrast, there I was, still free to choose my own path, and I was compromising almost before I began. What, for me, was worth doing? What would give meaning to my life? Was it string theory? Lattice theory? Or was it simply "fitting in" at a place like Caltech?

In his office, Feynman told me how he had found his place in life, in physics.

> I was supposed to be in physics. You know how I know? You see, I had a lab when I was a kid, and I used to play in the lab. I used to say I did experiments—but I never really did experiments. When I got to college I realized what an experiment really was. An experiment is a measurement to check some sort of idea. But that was not what my experiments were. My experiment was to make a photocell that rings a

bell when you walk in front of it, or to make a radio work or something like that. It wasn't an experiment to find out anything. It was just playing. I used to play in my lab. And I used to repair radios. In this town, in the Depression, and I was only a boy so it didn't cost so much . . . and I made myself a little kit, and bought parts. I understood what I was doing. I did enjoy very much, just making things.

Then I discovered this ability in theoretical analysis. At first I went to MIT as a freshman in the math department. I went to the head of the math department and asked, sir, "What is the use of higher mathematics if not to teach more higher mathematics?" And he answered, "If you have to ask that question don't go on in mathematics."

He was absolutely right. And that taught me something.

I had chosen mathematics only because I discovered I could do math very well. And I had somehow gotten the idea that math was at a higher level. But I really got interested in math because of application science. I hadn't fully appreciated that.

I was interested in math, and I was interested in all these things in terms of some kinds of use. And by use I meant application, understanding nature—DO something with it. Not just make more of this, this logical stuff, this monster. Of course, there is nothing wrong with it. I'm not trying to put down the mathematician. Everybody has different interests. But I realized that my interest is not in the precision of proofs, but in the thing that is proved, which is not the ordinary attitude of the mathematician.They like to structure the nature of proofs and so on. I was more interested in the facts that were demonstrated about the mathematical relationships. Because I wanted to use them for something, you see. So the attitude was different.

I found my place in physics. That is my life. For me, physics is more fun than anything else or I couldn't be doing it.

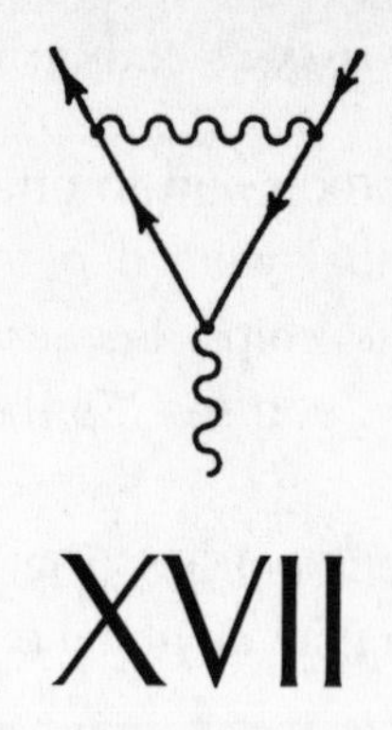

XVII

I STOOD IN MY KITCHEN and sipped strong, sweet, syrupy espresso. I had no inkling that it was the beginning of what would become the worst day of my life.

I was up early because a professor I knew from my undergraduate days was in town. He had been a kind of mentor to me, but I hadn't seen him in years. We were scheduled to meet at the Athenaeum for a late breakfast, or, as he called it, lunch. Afterward, he had a flight back to Boston and I had to run off to the doctor.

For me, in those days, "up early" meant around ten. Makes me sound like a slacker, but ever since my undergraduate days I had gotten used to working well past midnight. It's a tradition among physicists that goes back at least as far as René Descartes, in the seventeenth century. Descartes never got up before noon. He must have been a pioneer in this tradition because people didn't understand, and it earned him the reputation of being lazy. Still, he managed to revolutionize the fields of physics, mathematics, and philosophy. Not bad for a lazy guy.

As a graduate student I romanticized my work. I would sleep late, work late, and party hard. I might not revolutionize three fields, I thought, but at least in these respects I could be like young Descartes. Given my hours, and the fact that my thoughts and energies were dedicated almost exclusively to my work, I didn't have much contact with the outside world. Even the parties were mostly with other students. But I was content to feel connected to my peers, both contemporary and through the ages. To me, physicists separated in time like Einstein and Newton—and of course Descartes—were as much a part of my community as physics friends who lived elsewhere. We were all members of a noble society, each contributing whatever bricks he could to the edifice of theoretical physics.

Being on the faculty at Caltech, it was somehow different. The immersion wasn't there. When I studied string theory I found myself glancing at the clock far too often and seeking distraction whenever possible. I didn't connect much with my peers, but the night janitor was particularly friendly, so instead of late nights talking physics, I ended up learning quite a lot about professional soccer in Mexico.

What had kept me up late the night before had been the revival of an old diversion—writing. It had all started during one of our late-night *Hound of the Baskervilles* screening parties. As my neighbors and I watched, we would, as usual, yell out funny alternative lines of dialogue. And then it struck me—this was a film dying to be made fun of. So I started to write a parody of the film, along the lines of *Airplane,* a movie I had seen five times when it came out a year or so earlier.

Though I'd been writing short stories on and off since I was nine, I was too embarrassed to tell anybody at Caltech about the screenplay. Physicists, especially theorists, were often missionaries, or just plain snobs. Writing literature might be deemed barely acceptable, but a screenplay would definitely come in below zero on the lowbrow scale. I was supposed to be obsessed with physics, not Sherlock Holmes.

I thought about this as I arrived at 11:30 at the Athenaeum to meet my professor friend. We had been close in my undergrad days, and I wondered if I should ask his advice about both my research difficulties and my new interest. I wasn't sure how he'd react. When he showed up the first thing that struck me was that he looked exactly as I had left him—portly, avuncular, with bushy gray hair and a big beard. I even thought I recognized his sport jacket. The only novelty in his appearance was a crumb in his beard, presumably a leftover from breakfast and not my undergraduate days. I found it strangely endearing.

The waiter, a dressed-up student on work-study, brought us flat bread and butter. We sipped from our elegant water goblets and glanced at the menu. I didn't ask my former professor what he was working on—he had done some good work twenty years earlier but I didn't remember him publishing much while I knew him. But I did tell him I was looking at string theory. He knew it from its beginnings in the early seventies, but he was surprised to hear anyone was still working on it. In my mind I filed him in the camp of the oblivious, as opposed to the camp of the skeptics.

"Just be careful how you manage your career," he said. "You can't jump around too much from field to field, or you'll have trouble getting your next job. To establish your name, your research needs to have a certain coherence."

"Sometimes I think I'll never write another paper."

"It can take time. Don't panic."

"I'm not panicking. I'm more . . . discouraged."

"We all go through those times. It's part of the process."

"Maybe I'm not cut out for this," I said.

"Look, I believe in you. Hang in there."

"Thanks."

He chuckled. "What else would you do, anyway?"

"I haven't really thought about that."

"Of course not." The way he said it, I didn't know if he felt me incapable of anything but physics, or simply felt that nothing else existed.

"Well, I am doing some writing," I said finally.

"Writing?" He seemed puzzled, as if the only kind of writing he could imagine was practicing your penmanship. "What are you writing?" he asked.

"I've started a screenplay."

"What? You're writing a screenplay?"

He uttered this sentence with a strange cadence, as if he were my father, and he was saying, *You mean, this recent procedure you had was . . . a sex change operation?*

"Why on earth would you do that?" he said with sudden vehemence.

"I don't know. 'Cause I like it, I guess."

I looked down at the menu. This was getting uncomfortable.

I said, "The vichyssoise is really good here."

The scene felt surreal, and no lame attempt to change the subject was going to get me out of it, but, ever the optimist, I tried anyway.

"You know, we really should order. I have to run off to a doctor's appointment in a little while."

"Look," he said, "you owe it to yourself, and to me and to a lot of people, to keep at your physics. We put countless hours into your training. Years! You can't just throw it away like that. Your talent. Your schooling. It's an insult. A disrespect! And for what? Fiction? Worthless Hollywood crap?" His face turned red. The breakfast crumb fell from his beard.

I was caught off guard by his anger. On one hand, I had in no way meant to imply that I was thinking of giving up physics; on the other hand, I felt like saying *how dare you tell me what to do with my life*? Yet he had tapped into my feelings of being unworthy. Why *was* I working on such useless Hollywood crap? I tried to backtrack.

"I didn't exactly say I wanted a job in the movies."

"Why else would you write a screenplay?"

"It's just a hobby, that's all."

The student waiter came by.

"Just remember your responsibility. You have a talent. You have to make something of your life."

The waiter flashed me a knowing smile. He must have thought we were father and son.

I ordered the vichyssoise and an omelet. The professor had an omelet, too, but skipped the vichyssoise. Apparently he wasn't interested in culinary recommendations from an intellectual pervert. Halfway through lunch, a fresh crumb took up residence in his beard. We settled into generic small talk. I was relieved when the time finally came that I had to leave for the doctor, though that relief proved to be misguided.

With more perspective, I suppose I could have looked on Professor Breadcrumb's tirade with amusement. Stuck in his own narrow field, unable to appreciate the creativity of others. But I didn't have that perspective then, and his tirade really bugged me. Eventually, I talked to Feynman about it. And though he shared a certain scorn for much of modern literature, he respected the writer, just as he seemed to respect all endeavors that require the trait he admired most: imagination.

> I once thought about writing fiction for a little while, myself. Of course I've given lectures; that is to say I talked where they've been recorded. But that's an easy way out. So at a party at the English department I asked them, for the fun of it, how I would go about writing fiction, and this man who I respected very dearly, a professor, said, "All you have to do is write."
>
> I got ahold of the Grimm's Fairy Tales. I said they can't be very difficult to write . . . they can do whatever they want because they have angels, and trolls, and things like that. So they can do what they want,

there is all kinds of magic. So I said, "I'm going to make one of these up."

I could not make anything up but a combination of what I'd read. I felt unfortunately that when I recombined it, that I didn't have a deeply different plot, some cleverness, something different, some surprise, whereas the next story had some sort of surprise, not like the other stories. It had trolls in it again, but the nature of the plot, the twist was highly different. . . . And I said, "There's no more possibility here." And then I read the next one and it's entirely different. So I don't think I have the kind of imagination to make up a new story very well.

That's not to say I don't have a good imagination. In fact, I think it's much harder to do what a scientist does, to figure out or imagine what's there, than it is to imagine fiction, that is, things that aren't there. To really understand how things work on a small scale, or a large scale, it turns out it's so different than you expect, it takes one hell of a lot of imagination to see it! We need a lot of imagination to picture the atom, to imagine that there are atoms, and how they might be operating. Or to make the Periodic Table of Elements.

But the scientist's imagination always is different from a writer's in that it is checked. A scientist imagines something and then God says "incorrect" or "so far so good." God is experiment, of course, and God might say, "Oh no, that doesn't agree." You say, "I imagine it works this way. And if it does, then you should see this." Then other guys look and they don't see it. That's too bad. You guessed wrong. You don't have that in writing.

A writer or artist can imagine something and certainly can be dissatisfied with it artistically, or aesthetically, but that isn't the same degree of sharpness and absoluteness that the scientist deals with. For the scientist there is this God of Experiment that might say, "That's pretty, my friend, but it's not real." That's a big difference.

Suppose there was some great God of Aesthetics. And then whenever you made a painting, no matter how much you liked it, no matter how much it satisfied you, no matter what, even if it sometimes didn't satisfy you, anyway you would submit it to the great God of Aesthetics and the god would say, "This is good," or, "This is bad." After a while the problem is for you to develop an aesthetic sense that fits with this thing, not just your own personal feelings about it. That is more analogous to the kinds of creativity we have in science.

Also, writing, unlike math or science, is not one body of knowledge which is expanding and everything is put on together, a big monstrous being built by people together, in which there is a progress. Can you say, "Every day we're getting to be better writers because we've seen what has been written before?" That we write better because other guys have shown us how to do this or that earlier so now we can go on and carry it further? It is that way in science and math. For instance, I read *Madame Bovary,* which I thought was wonderful. Of course it was nothing but the description of an ordinary person. I'm not sure of my history, but I think *Madame Bovary* was in the beginning of writing a novel about ordinary people. I suppose that if other people's novels looked like that to me I'd be happy. But the modern novel, they're no longer done with that kind of craftsmanship, with that detail. The few that I've looked at, I can't stand them.

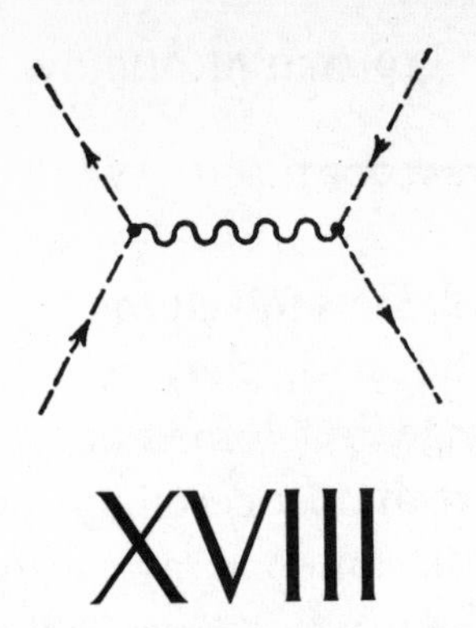

XVIII

MY DOCTOR WORKED at a small clinic in town. It wasn't far, so after my lunch with Professor Breadcrumb I walked over. The day was beautiful and sunny. The inside of the clinic was somewhere between sterile and tacky. Despite my appointment, I had a forty-minute wait to be seen. As I waited, I played in my head with ideas for my screenplay much in the way I often did with ideas in physics, so I didn't mind the wait.

The doctor was an older man, a little overweight. His face was round and inviting, like a smiley-face. It added to that image that he was almost completely bald. I was comfortable with this doc, which was a good thing since he had his hand on my testicles. I tend to be choosy about who I let do that. Especially if they are male.

"How long have they been like this?" he asked.

At first I thought it was a funny question.

"Like what?" I asked.

"These lumps?" he said.

Lumps? I was confused. What was he talking about?

"Here," he said. He showed me.

Technically, he said, they were just suspicious lumps at this stage, but lumps on your testicles that feel like this were almost certainly cancerous.

That was rare in someone my age. And I had one on each testicle, which was so rare, he said, it could be publishable. I thought I detected excitement in his voice. He was, after all, a past president of a prestigious professional society. But I was in such shock his remarks didn't offend me. All I could think was, this is not possible.

He told me the next step was a blood test to see if some hormone level was elevated. We ought to schedule a talk with a surgeon, he said. I felt the blood rush from my head. I collapsed into a seat. At this point his radar finally picked up that I was human, and not some poor clueless dog in his lab. He suddenly brightened a bit and to comfort me, I suppose, he told me that—if the cancer hadn't spread—after they removed my testicles hormone pills and a prosthesis would allow me to live a near-normal life. I wondered what Dr. Smiley-Face meant by "near-normal." To me, forget your pills and your voice goes up an octave is pretty far from normal. And how do you explain to your girlfriend those fake, nonfunctional testicles? No, I figured, life would never be "near-normal" for me again.

That was it. In an instant, my life had changed. My mother's mother had died of cancer at age forty. It was some kind of tumor that occurs between the bladder and kidney. They were wealthy, but this was in Poland in the 1930s, and there wasn't much to be done. It had apparently been a slow death, and excruciatingly painful. There was morphine, but it didn't help. My mother often recounted with tears hearing the screams of her mother every night. She told me of one night when she slept over at a friend's house, and how, when she came back, her father berated her for abandoning her dying mother that night, and for forgetting her family's pain. She never went out with friends again after that. Then her mother died. A couple of years later Hitler eliminated her family, her friends, and the need to balance their concerns. To this day my mother hasn't forgotten her family's pain. Neither had I. Even in my twenties, cancer had been my biggest fear.

It seemed to be the year for cancer at Caltech. Feynman dealt with his impending death by doing everything prudent to fight it, but also by proceeding mostly with calm acceptance. Murray had fought like crazy to save his wife, and his panic and sadness had both been apparent. How would I handle it? And how long would I last? I thought about all those times I had felt sorry for Feynman, while, it seemed, all the time it was I who would turn out to be the poor sap.

At first I walked around in a daze after getting the news. If I was unable to concentrate on physics before, afterward I was unable to concentrate on anything. I had trouble following simple conversations. Still, I went through the usual motions and told no one. Constantine took me aside and asked if I was on drugs. I think Ray just assumed. When I was alone, I felt sorry for myself. I cried often, and it would sometimes go on for what seemed like hours. After a few days, when my brain would work again, there was not one moment when my death was not in the foreground, along with the sinking feeling it produced in my stomach. Death became the focus of my life.

I looked at the olive trees on campus. Their beautiful craggy shape. Their pleasing gray color. Suddenly, everything seemed precious. The landscape, the sky, the elegant line formed in my apartment where the off-white wall met the cottage-cheese ceiling. I thought of Feynman staring at the rainbow. That was me now, desperate to appreciate all the little experiences of existence, even the ones that used to annoy me.

In a few days my doctor called. The blood test came back negative. The hormone level was not elevated. Relief. Ecstasy. But quickly shattered.

"That is often the case that the test is negative," he said. "It really doesn't mean anything."

I felt lost. Confused. I couldn't get a handle on what was going on.

"Why'd you take the test if it doesn't mean anything?" I said.

"It would have been the easiest way to confirm the diagnosis. But there are other ways. It's a formality, really."

"Will you take a biopsy?"

"No, we usually just remove the whole testicle."

"But this is both testicles."

"I'm afraid this kind of mass is always malignant," he said. I figured I was more afraid than he was. "We'll talk when you come in," he said. With that he ended the conversation. God hung up on me.

I felt lost. How did I let myself get into this situation? I had a Ph.D. in physics. According to a study I once read that meant, on the average, that I was about 25 percent smarter than Dr. Smiley-Face. But he was the expert. And I was left begging for his time and explanations. I decided to drive down to the USC medical school and educate myself, find a book and read up all about lumps and testicles. On the way I fantasized finding an array of benign explanations. Like cysts. Or bunions of the balls.

Unfortunately, testicles do not seem to suffer such fate. The books seemed to back him up.

When I got home I sat on my beanbag chair. Outside the heat of the day was dying down and the sun was low enough to be alluring instead of oppressive. The pool in the courtyard outside my door was deserted, except for a neighbor's cat crouching on the concrete beside it. As part of my new appreciation of life and nature, I watched the cat. How cute, I thought, the way it would crouch and pounce, practicing its long-lost ancient art of hunting.

Then I realized it wasn't practicing alone. The cat was toying with a young mouse it had caught. It would crouch, motionless, until the mouse tried to run away, then it would pounce and capture it. After a moment, it would let the mouse loose and repeat the game. Instead of drawing calm from gentle Mother Nature's beauty, I found myself receiving a depressing reminder that shit happens. It reminded me of Feynman and his multiple cancer surgeries. But if God was toying with Feynman, at least Feynman seemed to be enjoying the end of his days. I didn't think I could say the same for the poor mouse. Or me.

Ray came over.

"I can see there are dark clouds blanketing Mount Leonard," he said.

I still hadn't told him about the lumps but the dark clouds were impossible to hide. So I shrugged. He smiled.

"Don't worry," he said. "Dr. Ray brought medicine. Not exactly what's prescribed by the medical profession, but it'll do."

"Screw the medical profession," I said. "But I've been smoking too much." Suddenly I wondered if the marijuana smoking was related to the lumps.

"I need a light," he said, ignoring my response.

I got up and looked for some matches. He picked up a copy of a paper on string theory and flipped through it. Like most physics research papers, it was full of equations.

"It's theoretical physics but it looks like just math," he said.

"Call it purposeful math," I said.

"I hate math because of my dad," he said. "He was an engineer, rose out of the ghetto—we're talking Spanish Harlem, man—and goddamn it, he was going to make me an engineer, too. To him it was a matter of survival. To him, it was either learn math, or end up on welfare. So he'd test me on my arithmetic. And every time I got an answer wrong, POW! He'd hit me. And I mean hard, so I really felt it. No pussy-footing around from my father, no sir. What's nine times eight? POW! What's six times twelve? POW! That's why I hate it and that's why I'm good at it."

He lit his pipe and offered me some. I needed it badly.

"No, thanks," I said, and then regretted it.

"My dad should have forced me to smoke dope instead of doing math. Then I'd have grown up hating dope and loving math. Maybe I'd be a physicist like you. Not bad, cavorting with famous scientists, sleeping till noon. But what the hell, I like picking up garbage. I get off work early in the day, and I get to be outside." He looked over the research paper again. "I bet you have to really concentrate to do stuff like this."

"Yeah," I said. I felt I understood how he felt. I was both he and his father in one, forcing myself to study what I didn't want to, and beating myself up when I didn't get the answers fast enough.

He tried to hand me the pipe again. This time I took it.

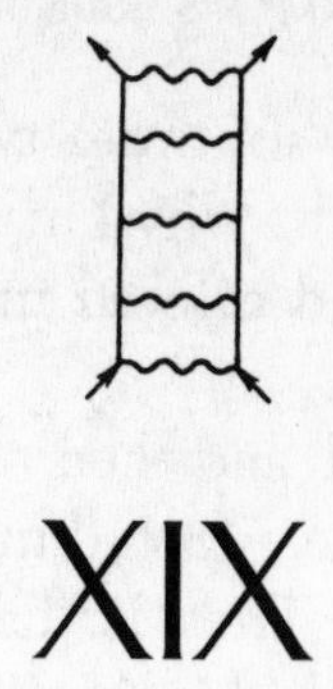

XIX

I WALKED TOWARD FEYNMAN'S office. My jeans had a rip at the knee and my flannel shirt was into its third day. But I didn't think about that. I was focused on the idea that Feynman and I finally had something in common. Impending death. Maybe we could form a support group of two.

I noticed Helen standing in her doorway, chatting with a student.

"Hello," she said as I approached.

"Hi," I said. I stopped at the mailboxes and pretended to sort through the two stale items of junk mail in the slot under my name. It was a stretch to linger there, but I didn't want Helen shooing me away from Feynman's door. Finally, her phone rang, and she disappeared into her office. I stepped quickly past. I knocked on Feynman's door. No answer. I knocked again.

"Yes," came his muffled voice from inside.

I opened the door and took a step inside. He was sitting on his couch, looking at a pad of paper he was holding. Finally, he looked up at me.

"I'm too busy to talk," he said. And when I didn't immediately move, he added, "Go away."

"I have a physics question," I said.

Of course, this wasn't true. But if I let on that my real purpose was personal, I'd never get in. And I certainly wasn't about to blurt out the whole truth, *I came to chat 'cause we're both dying of cancer.*

After a pause he said, "Not now."

His tone was softer now that he thought my visit was about a real physics question.

"Okay, when's a good time?"

"I don't know. Try me next week."

Next week wouldn't do. Next week I might be dead.

I said, "Okay." I stepped back. "Anyway, it was a long shot that you could help. It's a question about quantum optics, and I'm sure you haven't thought about that subject in years."

A good friend of mine from graduate school named Mark Hillery had gotten a position in New Mexico doing research in quantum optics. We had been talking over the phone about his work and mine on and off amid my study of string theory, mainly on nights the janitor was too busy to provide a diversion. Like my writing, my dabbling in quantum optics was not something I shared with my colleagues. It'd be considered lowbrow. Too applied. But Feynman appreciated all aspects of physics. And he always liked a challenge.

I started to close the door. Slowly.

As I just about had it shut he said, "Wait."

He was curious now and, most of all, wanted to

show me that there was no problem in the world of physics into which he couldn't provide the greatest insight.

"What's the problem?" he asked.

My ploy had worked. Now I had to come up with a question. That wasn't hard.

One of the major issues in quantum optics was to describe how beams of laser light behave when they penetrate a material such as a crystal. Due to the presence of the material medium, they behave quite differently than when they propagate in a vacuum. Mark and I had found that we could use the methods of my dissertation—approximation by infinite dimension—to model the individual atoms inside certain crystals, and, with some assumptions and a lot of mathematics, develop a theory of how laser light and the crystal interact.

There already existed a theory describing those interactions, but it wasn't derived from a theory of the individual atoms, as ours was. Instead, it had been derived by approximating the crystal lattice of atoms as a continuous medium with certain macro-properties that were measured by experiment. If the crystal were a cup of water, then the old approach would be to treat the water within the cup as a liquid with certain macroscopic properties, such as density, viscosity, and refractive index (a measure of how it bends light), and to ignore the fact that it was actually made of microscopic things called water molecules. Our approach was to start with the water molecules, and then derive everything else. If we could really "derive" everything else, then because we didn't ignore the "details," ours would be a clearly superior approach. But to do what we wanted was a far more complicated undertaking than the old approach, so in order to carry it out we had to make our own simplifying approximations. The central one was to employ my method of infinite dimensions. Since both the old way and our way thus involved approximations, neither was inherently a better method. Still, we thought redoing the theory our way might lead to some new insights about the physics. Like Feynman's work on liquid helium, this theory would be a model created for a given situation, not a fundamental theory like quantum chromodynamics or string theory. But it seemed interesting, so we worked on it.

Mark compared our theory to the usual theory and called one night to report that they didn't agree. I looked up the fifteen-year-old paper in which the old theory was first presented, and sure enough, our results, though similar, had a major conflict. Obviously, one or the other of the theories was wrong, and we figured it was us. Somewhere we had either made a mathematical error or made an assumption that to not justified. I figured finding it would be a great problem to discuss with Feynman.

Feynman grasped the idea behind our theory immediately, proving to me that there indeed was no problem in the world of physics into which he couldn't provide the greatest insight. In fact, in the next half hour he provided me with more insight into it than I had had in the two months I had been thinking about it. I should have been discouraged by the ease with which he surpassed my own thinking, but instead I was excited because he liked our idea.

Then I told him about the conflict with the other theory.

"Do you understand their theory?" he said.

"I read the paper. I followed most of it."

"Followed it? Just because you are following someone doesn't mean you are going down the correct path. When you can derive it yourself," he said, "then you understand it. And maybe you can believe it." After a moment he added, "Of course, you might find that it is bullshit. I suspect it is, because it looks to me like you did everything correctly."

"But the theory has been around for fifteen years," I said.

"Okay," he said. "So not only is it bullshit, it is old bullshit."

I laughed.

We never did get around to talking about our impending deaths, but it was a support group nonetheless. For the short period of our conversation, I was provided with an escape from my constant worry about cancer. When we were talking quantum optics, the world seemed wonderful and exciting. I got the impression Feynman felt the same way.

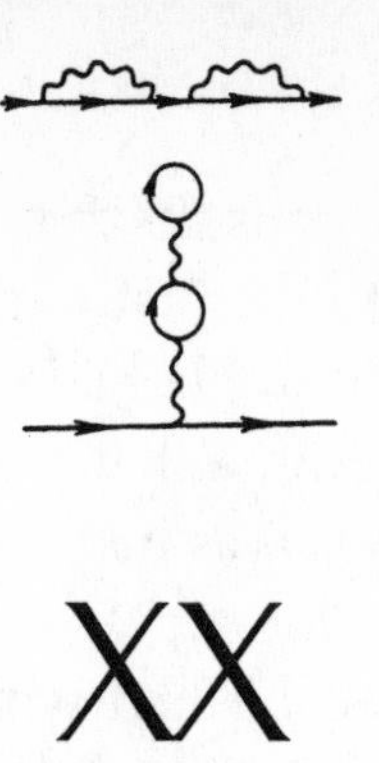

XX

IT WAS TIME to see Dr. Smiley-Face again. As I approached the clinic, my stomach tightened up. By the time I got there, I must have looked pale and awful, because this time they didn't make me wait. I was immediately shown to an examination room, and told I could lie there and rest if I liked. Yeah, they treat me nice, now, I thought. Because they feel sorry for me.

Lying on the papered cushion, I pictured all the nasty procedures that might be in my future. The surgery, of course, which in itself was too terrible to contemplate, and then the endless tests, injections, X-rays, and maybe radiation or chemo, which meant further internal mutilation. Terrible nausea, and losing all my hair, even my eyebrows and lashes.

After a few minutes my doctor opened the door. I sat up, suddenly feeling adrenaline rush into my body. He seemed surprised that I was alone. He started to retreat from the room.

"Doctor?" I said.

"I've asked for a consultation," he said. "The best people we've got. It'll be just a minute."

Then he stepped out. He had sounded grim. I wondered what this meant. What was in store for me? I felt shaken. The worst was just not knowing what was going on. I lay back down.

When he returned he had with him, not one, but two specialists, a testament, I figured, to his excitement over my illness. He was showing me off. Soon, three serious men stood there, huddled over my balls. Unlike physicists, these doctors did wear white coats. For some reason, it made the whole episode even scarier. As if they wore them to insulate themselves from my blighted body.

One specialist muttered something to the other. They both nodded.

The second specialist left, and the first one looked at me.

"You have lumps," he told me, "but they aren't cancer. They aren't even tumors. You are just fine."

I looked at him, and for a moment I was relieved. My whole body relaxed as if I had been given some sort of injection. Tears came to my eyes, then streamed down my cheeks. I looked at Dr. Smiley-Face. Then suddenly I was thinking, you said the lumps were malignant. Why do these other guys think the lumps are okay? Do they have X-ray fingertips? What kind of medicine do you practice, majority rules?

Dr. Smiley-Face answered the questions that must have been apparent on my face.

"The lumps are the same on both sides," he said.

"They're mirror images," the specialist said. "Tumors would not grow like that. So you must have been born that way. You are fine. Hasn't any other doctor ever remarked on them before?" No, the landscape of my testicles had until now been virgin territory.

Dr. Smiley-Face apologized, and that, for them at least, was that. For me, for years after that incident, I had trouble believing Dr. Smiley-Face was not actually right. Articles in the newspaper on testicular cancer would give me a sinking feeling in my stomach and the blood would drain from my head and I'd have to sit down to avoid fainting. I would get strange looks from the doctors I saw for unrelated ailments, as I asked them to, by the way, also check my testicles.

I'm finally over that. I figure, if it had been true, I'd be long dead by now. The problem with my genitals was congenital. I had been saved by symmetry.

XXI

I DROVE BACK TO the apartment so elated I almost had two serious accidents on the way. I thought about the irony of dying just after learning I wasn't dying. I thought, you don't need cancer to die. It could come just like that, from a moment of carelessness. You get into the car. You are terminally ill, but you don't even know it until that last moment when you're slamming on the brakes.

I tried to get hold of myself, but after seeing the doctor I felt a definite high. There must be some kind of elation hormone your body releases. You could get rich if you packaged it, but they'd probably make it illegal. It sure made it hard to focus on the road. It must have affected my psychology, too, for now that my ordeal was over, and I should have had less need to talk, I suddenly felt like telling someone what I had just been through.

I started with Ray. I found him strolling out by the pool, freshly showered after his day slinging trash. As he listened, his face contorted into a series of expressions, as if within seconds he had gone through all the stages of grieving—shock, denial, anger, depression, acceptance—and then relief. He grabbed me in a powerful hug. Pressed close against him, I felt the stubble of his beard like fine sandpaper on my cheek. I could smell talcum, mixed with just a touch of the lingering, sour odor of garbage. When he released me all he said was "I'm glad you're okay."

We decided I should take the next few days off. Ray, too. Well at least one day. We partied late into the night. The next morning he called in sick, sick with joy for me, and we kept the party going. We had everything we liked, a kind of celebration of life. That meant pizza for breakfast, burgers for lunch, and pizza and burgers for dinner. Plus plenty of joints and beer and cigars in between.

Late in the afternoon Ray dropped his own bombshell. He was leaving. Moving to Bellevue to be with his new love, the Microsoft woman. She said he could live with her for a while up there before he needed to get a job, so he was thinking of giving up the garbage man business and learning how to program computers. Finally putting his mathematical talent to some use. Time to stop punishing himself along with his father, I guess.

Funny how quickly my elation bubble could burst. I was already lonely, and the thought of the person who had become my closest friend in town disappearing made me reel. I should have been happy for him, but it felt like another punch in the gut.

By the next morning our marathon party had made both Ray and me ill. Ray called in sick again, this time legitimately. And I spent the day in bed chewing aspirin and sipping tea, pondering the question, *Now that I have my life back, what should I do with it?*

It was sweltering outside, "unseasonably warm," as they put it on the radio. Maybe, but it was a reminder that summer was near. The academic year would soon end. I thought about what I had done, and not done. I hadn't accomplished much. No great discoveries, not even any publishable work, unless Mark and I figured out our optics theory. But I was still alive. I thought back on my talks with Feynman. To me life and career had seemed very complicated. He made it all sound simple. If an ape could do it, so could I, he had said. But I wasn't an ape. I worried about how it would all turn out. I figured apes probably didn't do that. Is that what you learn as you grow older, that it's all not as complicated or important as you thought?

When I got back to Caltech, I found I had missed some big news. It concerned Constantine. We had never spoken again about the possibility of my working with him. Now his postdoctoral appointment was ending and he had lined up a new job in Athens starting the next fall. That was news but that wasn't the big news.

Constantine's claim to fame was his computer calculation of the mass of the proton from the theory of quantum chromodynamics. Now there was a rumor going around: that Constantine did not translate the problem to the computer in an honest way. There is not a unique way to translate equations from the real continuous space of the mathematical theory to a finite lattice of points that the computer can handle, so lattice theory is as much an art as a science. You try to follow accepted principles regarding what makes the best sense in terms of reliability and accuracy. And then you let the computer grind through it. Work in lattice theories is harder to check than purely mathematical work, because though you can follow how the problem is set up, you cannot mentally go through all the steps the computer makes while performing a calculation. According to the scuttlebutt, Constantine had worked backward, knowing what the proton mass was and playing with the parameters of how he set up his particular calculation in order to get the right answer. It is a subtle difference, perhaps, but it is important to disclose.

Constantine wasn't denying it. And he pretended not to care about the fuss. He just waved his arms and dismissed it with the same all-knowing confidence he had when he discussed Greek or American politics. "What's the big deal?" he said. "I used what I knew to improve my computer model. Everybody does that." But he puffed constantly on his cigarettes. Short, joyless puffs.

I felt bad for him, but I was also a little angry. He was my good friend, and I had trusted him. I still felt he was trustworthy on a personal level, but it would be hard to ever have the same respect for him again. I didn't tell him about my cancer scare.

But I did want to tell Feynman.

I used my mailbox trick to make sure Helen didn't see me, and then I burst into Feynman's office with only a perfunctory tap on the door. He was resting on the couch, not working, and didn't seem to mind the interruption.

To break the ice I mentioned the Constantine controversy. He just shrugged.

"I haven't read his paper. I don't know enough about it. What do you expect me to say?"

"I thought you'd say something like, what a louse! He did it because he thought what was important was success, not discovery."

"Hell, no. I'm not going to psychoanalyze the guy. But what should bother you as much as whether or not your friend fudged his work, is that a lot of people read it and couldn't tell the difference. There are so many people out there not being skeptical, or not understanding what they are doing. They're all just following along. That's what we have—too many followers, too few leaders."

I sat down. I'd had enough about Constantine. I wanted to talk about me. I told Feynman my cancer story.

He shook his head. "At least a stupid physicist hurts no one but himself," he said. "You know, I had a lot of doctors who told me they couldn't operate on me. But then I found the one doctor in the country brave enough to try it. It was a very long operation. Very thorough. Of course, chances are he may have missed some. You can't know. We'll just have to see."

He shut his eyes.

I gazed at him. He looked drained this day, his face pale, thin, and wrinkled. For the first time I saw him, not as a physicist, or a legend, or as my sometime pal down the hall, but merely as an old man.

He opened his eyes. I was staring at him.

"You're thinking I don't look so good," he said.

"No, you look fine," I lied.

"Don't bullshit me. And you know what?"

"What?"

"You don't look so good either."

I smiled. "I've had a rough couple of weeks." I decided to leave out the part about the two-day party.

He cracked a slight smile. "With maybe some exhausting celebration at the end?"

I smiled back. "Yeah, a little. With Ray. Remember him?"

Feynman shook his head. He had obviously liked Ray. Somehow we got to talking about how Ray's father browbeat him into hating math.

"My son Carl and I," he said, "we love talking math." He brightened, as if a pulse of energy had infused him. "And he's very good."

"My dad and I never talked math," I said. "He never got past high school. The Nazis saw to that. But I always loved doing math problems. I like thinking hard. And I like the feeling you get when you figure something out, or create a new idea."

"Well then, that's the answer you've been searching for, isn't it?"

"What do you mean?"

"When I was talking to Ray he said he asked you why you liked physics and you couldn't tell him."

"Oh yeah." I was a little embarrassed that Ray had shared this.

"Well, you've figured it out. You like it because you

like thinking hard, you like being creative, and you like solving problems."

"I don't think that's the answer," I said.

"What do you mean *you* don't think that's the answer? That's not my answer. That was your answer." He sounded impatient. It's how he got when you weren't quick enough on the uptake. I tried to explain myself.

"Okay, I said that, but it couldn't be why I like physics, because it's not really specific to physics."

"So?"

"So it applies to a lot of pursuits."

"So?"

With that, Helen peeked in. "Professor Feynman, is he bothering you?" She turned to glare at me, but continued speaking to him. "I know you were trying to get some work done."

"It's okay, Helen," he said. "He wasn't bothering me." Then, to me, "But he's starting to."

"Then it looks like I came just in time," Helen said. "Come on, Dr. Mlodinow. I noticed that after lingering at the mailboxes, you neglected to pick up your mail after all." She held it out for me. So much for my ruse.

"Give me just one more minute, okay, Helen?"

She sneered, but Feynman didn't object, so she left. I turned to Feynman.

"I think I see your point."

"All right."

"The term's ending soon, so . . . in case I don't see you before the summer . . . I just wanted to thank you . . . for all you've taught me."

"I haven't taught you anything," he said.

"You've taught me about myself."

"That's bullshit. What have I taught you?"

"I guess I'm still sorting it out . . . but like just now . . . you've taught me a way of looking at the world, I guess. And where I fit in."

"First of all, like 'just now,' I didn't teach you that, you did. I can't teach you how you fit in, you have to discover that yourself. And secondly, I'm a lousy teacher, so I doubt I have taught you anything."

"Okay, then . . . thanks for all the . . . conversations we've had. Whether or not you've taught me anything, I've enjoyed them."

"Look, if you're going to insist that I've taught you something, I guess I should give you a final exam."

"Really?"

"One question."

"Sure."

"Go look at an electron microscope photograph of an atom, okay? Don't just glance at it. It is very important that you examine it very closely. Think about what it means."

"Okay."

"And then answer this question. Does it make your heart flutter?"

"Does it make my heart flutter?"

"Yes or no. It's a yes or no question. No equations allowed."

"All right, I'll let you know."

"Don't be dense. I don't need to know. You need to know. This exam is self-graded. And it's not the answer that counts, it's what you do with the information."

We locked eyes. His younger face flashed in my mind. The energetic, smiling bongo drum player I had seen pictured in the front of his book, *The Feynman Lectures on Physics*. A question popped from my lips.

"Do you have any regrets?" I said.

Feynman didn't snap back that it was none of my business. He didn't do anything for a moment. I wondered if he would open up about his frustration with quantum chromodynamics. But then his eyes welled up with tears.

"Sure," he said. "I regret that I might not live to see my daughter, Michelle, grow up."

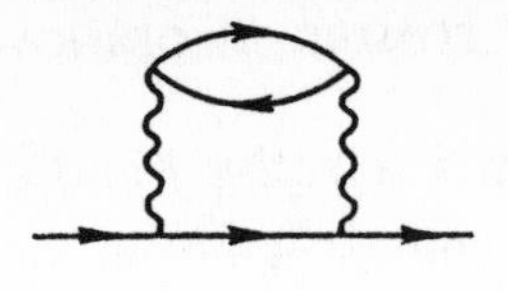

XXII

Of all the questions I had put to Feynman, the one that always stuck with me the most was the final question: Who are you as a person—and how has being a scientist influenced your character?

He hadn't liked the question—it was too psychological.

But he answered it.

Given his impatience with all questions psychological, I considered his answer a special gift. A notice to me that whatever monumental importance I might attach to success, in the end it is not success that really matters.

> I don't even know what that means, to understand yourself on a personal level. I hear people talking about things like, "I have to find out who I am." I don't know what they are talking about. I can say that certainly I've learned an awful lot about myself by studying biology. I know how I am put together. I have a big theory about how I operate mechanically. But that's not understanding yourself on a personal level.

I can say I am a scientist. I find excitement in discovery. The excitement is not in the fact that you've created something, but that you've found something beautiful that's always been there. So scientific stuff affects every part of my life. And affects my attitudes toward many things. I can't say which is the cart and which is the horse. Because I'm an integrated person and I can't tell you whether for instance my skepticism is the reason I'm interested in science or my science is the reason I'm skeptical. Those things are impossible. But I want to know what is true. That is why I look into things. To see and to find out what is going on.

I'll tell you a story. When I was thirteen I met a girl, Arlene. Arlene was my first girlfriend. We went together for many years, at first not so seriously, then more seriously. We fell in love. When I was nineteen we got engaged, and when I was twenty-six we got married. I loved her very deeply. We grew up together. I changed her by imparting to her my point of view, my rationality. She changed me. She helped me a lot. She taught me that one has to be irrational sometimes. That doesn't mean stupid, it just means that there are occasions, situations, you should think about, and others you shouldn't.

Women have had a great influence on me and have made me into the better person that I am today. They represent the emotional side of life. And I realize that that too is very important.

I'm not going to psychoanalyze myself. Sometimes it is good to know yourself, but sometimes it isn't. When you laugh at a joke, if you think about why you laughed, you might realize that, after all, it wasn't funny, it was silly, so you stop laughing. You shouldn't think about it. My rule is, when you are unhappy, think about it. But when you're happy, don't. Why spoil it? You're probably happy for some ridiculous reason and you'd just spoil it to know it.

With Arlene, I was happy. We were very happily married for a few years. And then she died of TB. I knew she had TB when I married her. My friends, they told me, don't marry her, that since she had TB I didn't have to marry her anymore. I wasn't marrying her out of a sense of duty. I did it because I loved her. What they were really afraid of was that I'd catch it, but I didn't. We were very careful. We knew where the germs were coming from and we were very careful. It was a real danger, but I didn't catch it.

So for instance, science has an effect on my attitude, say, to death. I didn't get mad when Arlene died. Who was there to be mad at? I couldn't get mad at God because I don't believe in God. And you can't get mad at some bacteria, can you? So I had no resentment and I didn't have to look for revenge. And I had no remorse because there was nothing I could have done about it.

I'm not worried about my own future in heaven or hell. I have a theory about that that I believe does come from my science. I believe in scientific discoveries and therefore have a view about myself that is consistent. Now I've just been to the hospital and I don't know how long I have to live. It happens to all of us sooner or later. Everybody dies. It's just a matter of when. But with Arlene I was really happy for a while. So I have had it all. After Arlene, the rest of my life didn't have to be so good, you see, because I had already had it all.

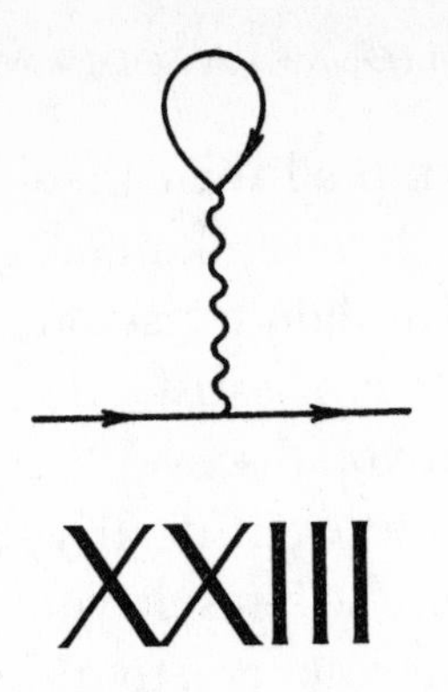

XXIII

WHAT IS IMPORTANT IN LIFE? It is a question we should all give thought to. The answer is not taught in school, and it is not as easy as it may seem, for a superficial answer is not acceptable. To discover the real truth you have to know yourself. Then you have to be honest with yourself. Then you have to respect and accept yourself. For me, these were all tough tasks.

I had gone through college and into academia in a hurry, wanting to rush ahead with my work, to prove to the world that I had been alive, and that it had mattered. That was an external focus to life. That was Murray's way. To accomplish and impress. To be an important person, and a leader. It was the classical path. The traditional one. It seemed to be an obvious and worthy goal. I had accepted it without second thought. But for me, it was like chasing a rainbow. Even worse, it was like chasing other people's rainbows. Rainbows whose beauty I didn't really see.

Through Feynman I saw another possibility. And just as the discovery of the quantum principle caused physicists to revamp all their theories, Feynman's example caused me to rethink mine. He didn't seek the leadership role. He didn't gravitate to the sexy "unified" theories. For him satisfaction in discovery was there even if what you discover was already known by others. It was there even if all you are doing is rederiving someone else's result your own way. And it was there even if your creativity is in playing with your child. It was self-satisfaction. Feynman's focus was internal, and his internal focus gave him freedom.

Our culture is a culture that, by Feynman's characterization, is Greek. It is a culture of logic and proof, rules and order. In our culture people who live their lives like Feynman are considered eccentric, for Feynman was a Babylonian. For Feynman, both physics and life were ruled by intuition and inspiration, and a disdain for rules and customs. He ignored the conventional methods of physics, and invented his own, his sum over paths and his Feynman diagrams. He also ignored academic culture and invented his own, eating with the students in the Greasy, or working on his physics in strip clubs, or doing research less for reasons of ambition than for reasons of love. And if his behavior was not approved of, well, what did he care what other people thought?

I chose Feynman's way. A lot of people aren't lucky enough to feel a passion for any particular endeavor, or else, like my immigrant father, are too occupied with mere survival to have any choices. Especially after my death scare, if I had a choice, I did not want to squander it. I resolved that, as long as I could, I would spend my limited time of life pursuing goals that moved me, whether or not others found them worthy. I resolved to never lose sight of the beauty in physics—and life—whatever that beauty is, personally, for me.

I knew I would have to take some risks because I would not stick to one narrow, "coherent" field of research, or even to a single career. I knew that since I was not driven by ambition I might not be accepted by my peers who are. I knew that I might be looked upon with the same misplaced scorn with which I had looked upon Professor Gardening, or with which Professor Breadcrumb had looked upon me. And I knew that, in the end, I would probably not find the success in conventional or material terms that Feynman achieved, or that my mother desired for me—or that Murray seemed to want to thrust upon his daughter Lisa. But at least with an internal focus my happiness would be under my own control.

Once I shed the burden of the real or imagined values and expectations of others, it was easy to tell where my passions lay. I dropped string theory. I started to work more on the quantum optics I had started with Mark. As it turned out, Feynman had been correct—our theory was right and the accepted approach was flawed. I also came out of the closet about my writing. If Feynman could see beauty as the inspiration for the theory of the rainbow, and if an electron could behave like a wave, and light like a particle, then the little contradiction of Leonard flitting among different subfields of physics, or even among varied careers, would not shake the universe.

Other than Feynman, none of my colleagues at Caltech took an interest in my optics work. And most of them rolled their eyes whenever I brought up writing. Before long, I was asked to move out of my office, to another one on the other side of the building. "Murray wants the office next to him for one of his own people," Helen said. I wondered about a connection with my new choice of activities, but, mainly I thought, *Who cares?* I didn't know where my physics or my writing would take me. But I looked forward to the ride. And whether I continued to write as a hobby, or ever supported myself with it, I hoped that maybe someday I'd write something that Feynman would admire. And then I thought, no, even better, I hope that someday I will write something that I will admire.

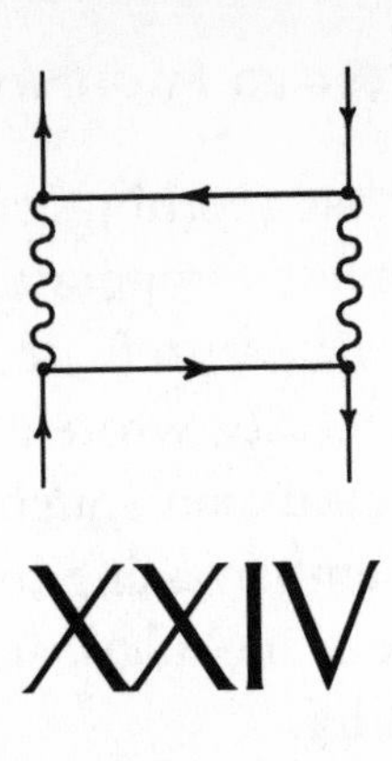

XXIV

AFTER I LEFT CALTECH I never saw Feynman again, except on television.

It was in early 1986. He was weak from his long battle with cancer, but he had nonetheless agreed to be the single scientist on the American presidential commission investigating the crash of the space shuttle *Challenger.* Impatient with the bureaucratic process, he flew around the country conducting his own mini-investigation. He soon zeroed in on a prime cause of the disaster that might have remained a mystery if not for his muckraking: the loss of resiliency of one of the shuttle's key gaskets, the rubber O-rings, at low temperatures. At the commission's televised public meeting on February 11, 1986, Feynman dipped an O-ring into a glass of ice water and showed that, when squeezed, it did not show resilience. With this now-famous, simple, desktop demonstration Feynman showed that responsibility for the disaster lay largely with NASA managers who ignored their engineers' warnings to abort the launch because of the unusually cold temperature that morning, twenty-nine degrees Fahrenheit (the lowest temperature at any previous launch had been fifty-three degrees Fahrenheit). Feynman, now a celebrity, wrote a report on his findings, which the commission sought to suppress, as it was thought to be embarrassing to NASA. But Feynman fought to have it included, and in the end it appeared as an appendix.

Feynman battled his cancer through two more operations, in October 1986 and October 1987. After the latter operation, his fourth, he had trouble bouncing back. He was now weak, in pain, and often depressed. But physics still brought him vigor. He continued to teach a course in quantum chromodynamics. And, in his last months of life, he finally decided to learn string theory. Murray taught him, in a private "seminar" they held each week.

On Wednesday, February 3, 1988, Feynman entered the UCLA Medical Center in Los Angeles. He didn't know the seriousness of his malady when he entered the hospital, but he soon found out. He had one remaining kidney, and it was failing. His doctors offered continuing dialysis, but that wouldn't provide much quality of life. It was not a path he wanted to follow. He refused the procedure. He accepted morphine for the pain, and oxygen, and prepared himself for the consequences. He said he saw it as his final discovery: what it is like to die. He told a friend he had realized when he was seven that it would happen someday, and he didn't see any reason to start complaining about it now. He said he would find the experience IN-ter-ES-ting.

Life gradually drained from him. First he couldn't speak. Then he couldn't move. And finally, he could no longer breathe. He had made his final discovery. It was February 15, 1988, just a few months before his seventieth birthday. He had survived his cancer for ten years, significantly beating the odds he had looked up so long ago. And he had held on long enough to overcome his greatest regret—he had lived to see his little daughter Michelle reach adulthood.

Six weeks after Feynman's death there was a memorial service for him at Caltech, a festive celebration of his life, with speaker after speaker coming to the stage to reminisce. Murray's name was on the program, but he didn't show up.

He had a good excuse.

As Murray was preparing to leave for the service, federal agents wearing flak jackets and carrying assault rifles raided his home. It turned out his interest in ancient cultures—and their artifacts—had led him to purchase some that had been smuggled into this country. Murray forfeited some artifacts, cooperated with the U.S. customs agents, and, in the end, flew to Peru where he was honored for setting a good example and given a key to the city of Lima.

Murray finally had the opportunity to pay public tribute to Feynman in a special memorial issue of *Physics Today* honoring Feynman. In his obituary, Murray wrote what can only be categorized as a "mixed review" of Feynman's personal style. It raised a few eyebrows in the physics community.

"What I always liked about Richard's style," Murray wrote, "was the lack of pomposity in his presentation.

I was tired of theorists who dressed up their work in fancy mathematical language or invented pretentious frameworks for their sometimes rather modest contributions. Richard's ideas, often powerful, ingenious and original, were presented in a straightforward manner that I found refreshing. I was less impressed with another well-known aspect of Richard's style. He surrounded himself with a cloud of myth, and he spent a great deal of time and energy generating anecdotes about himself. . . . Many of the anecdotes arose, of course, through the stories Richard told, of which he was generally the hero, and in which he had to come out, if possible, looking smarter than anyone else. I must confess that as the years went by I became uncomfortable with the feeling of being a rival whom he wanted to surpass; and I found working with him less congenial because he seemed to be thinking more in terms of "you" and "me" than "us." Probably it was difficult for him to get used to collaborating with someone who was not just a foil for his own ideas. . . ."

Murray and Feynman were rivals. Nevertheless I was surprised that Murray chose to be so harsh. That's Murray, still competitive, still tormented. But I prefer to think the real reason for Murray's negativity was, simply, that when Murray wrote the obituary he was having a bad day. In any case, I don't think Feynman would have been offended—he always appreciated it when you spoke your mind. Ironically, around the time Murray was writing the critical article, he was doing new landmark research based on Feynman's early work on the formulation of quantum theory in terms of paths or histories. Shortly after completing that work, Murray left Caltech. He now lives and works in Santa Fe, New Mexico.

By the time Murray left Caltech, John Schwarz no longer needed him as a mentor, for in 1984, Schwarz and Michael Green had a historic breakthrough. After working on the problem for five years, they found the mathematical miracle they were looking for and resolved the last major inconsistency in string theory. It didn't make the theory any easier to solve, but it convinced many leading physicists—especially Edward Witten—that the theory had too many miraculous properties to ignore. As Holmes, or more probably Rockford, might have said, *Coincidence? I think not.* Within months string theory, the laughingstock of physics, became string theory, the hottest thing in physics.

Over the next two years hundreds of particle theorists jumped on the bandwagon, writing over a thousand research papers. Today string theory research dominates the field of elementary particle theory. As rare as it had been to find anyone working on string theory, it has become just as rare to find a particle theorist not working on it. By the end of 1984, Murray was finally able to get Schwarz "a real job," as a professor at Caltech. But it still wasn't easy. As one administrator remarked, "We don't know if this man has invented sliced bread, but even if he has, people will say that he did it at Caltech, so we don't have to keep him here."

In 1987 Schwarz received a prestigious MacArthur fellowship, and in 1997 he was elected to the National Academy of Sciences. In 2001, he was awarded the

American Physical Society and American Institute of Physics' 2002 Dannie Heineman Prize for "valuable contributions made in the field of mathematical physics." Despite the glory, string theory is still a work in progress, far from proven or even well understood. Schwarz says he never had any regrets, even when it looked as if his work might never be accepted. He also says he never had any doubt it was correct. Today Schwarz has Feynman's old office, and still works on string theory. What is not yet known is how he will fare without the help of Helen Tuck, who, well into her seventies, has just retired as department secretary.

Feynman was not a fan of string theory, but he respected Schwarz. And why not? If anyone wasn't following the crowd, it was John. Whenever I hear people's ideas easily dismissed, or hear someone's goals in life criticized as unattainable, I always think of John Schwarz. And I think of Feynman, for if there is one thing he taught me, it is the importance of being truly committed to whatever it is we are striving for.

One day a year or so ago I was going through musty boxes that I had stored in a warehouse far out of town. In one of them, amid decades-old college texts, I found the cheap old Radio Shack cassette tapes which were to form the basis for the transcriptions in this book. When I recorded our conversations I didn't know I wanted to write a book, or even that I was capable of doing it, but I did know I wanted to write about Feynman. I imagine anyone who ever knew him, and had the inclination to write, would have felt the same way. Yet I did not write about him, and the tapes lay dormant for some twenty years. I think the reason was that, back then, I didn't really have any purpose in mind.

Listening to the tapes again after all those years, I missed Feynman, the gruff, reluctant teacher whose spirit even terminal cancer couldn't dampen. And I missed the person I was, the eager, innocent student with his whole life before him. It was then that the purpose of this book became clear.

In his epilogue, Feynman stated his own goal in writing *The Feynman Lectures on Physics,* which I had read on the kibbutz in Israel so many years ago. Feynman wrote, "I wanted most to give you some appreciation of the wonderful world and the physicist's way of looking at it." His statement was overly modest, for the worldview he imparted in those books was not just any physicist's way of looking at the world; it was distinctly his own. It is this goal that I hope I have furthered in writing this book. For Richard Feynman always knew how to get the most out of what the world had to offer, and how to get the most out of the talent with which God—or mere genetics—had blessed him. That's all we can hope for in life, and in the years since he's passed on, I've found it to be a valuable lesson.

Further Reading

By Feynman:

Richard Feynman, *The Feynman Lectures on Physics*
Richard Feynman, *The Character of Physical Law*

About Feynman:

James Gleick, *Genius*

With more emphasis on technical content:
Jagdish Mehra, *The Beat of a Different Drum*

By Murray:

Murray Gell-Mann, *The Quark and the Jaguar*

About Murray:

George Johnson, *Strange Beauty*

On string theory:

For the general audience:
Brian Greene, *The Elegant Universe*
F. David Peat, *Superstrings and the Search for the Theory of Everything*

If you have an advanced degree in mathematics or physics:
Joseph Polchinski, *String Theory*
Michio Kaku, *Introduction to Superstrings and M-theory*

LEONARDO MLODINOW received his Ph.D. from the University of California at Berkeley. He was on the faculty of Caltech, and an Alexander von Humboldt Fellow, before becoming a writer in Hollywood for *Star Trek: The Next Generation* and other hit television series. His first book, *Euclid's Window*, a critically acclaimed history of geometry, has been translated into eight languages. He lives in South Pasadena, California.